Marine Geotechnical Engineering and Marine Civil Engineering Construction

Marine Geotechnical Engineering and Marine Civil Engineering Construction

Guest Editors

Jun Hu
Guan Chen
Yong Fu

MDPI

Basel • Beijing • Wuhan • Barcelona • Belgrade • Novi Sad • Cluj • Manchester

Guest Editors
Jun Hu
School of Civil Engineering and Architecture
Hainan University
Haikou
China

Guan Chen
Earth Observatory of Singapore
Nanyang Technological University
Singapore
Singapore

Yong Fu
Department of Ocean Science and Engineering
Southern University of Science and Technology
Shenzhen
China

Editorial Office
MDPI AG
Grosspeteranlage 5
4052 Basel, Switzerland

This is a reprint of the Special Issue, published open access by the journal *Sustainability* (ISSN 2071-1050), freely accessible at: https://www.mdpi.com/journal/sustainability/special_issues/Marine_Geotechnical_Engineering_Marine_Civil_Engineering_Construction.

For citation purposes, cite each article independently as indicated on the article page online and as indicated below:

Lastname, A.A.; Lastname, B.B. Article Title. *Journal Name* **Year**, *Volume Number*, Page Range.

ISBN 978-3-7258-7771-3 (Hbk)
ISBN 978-3-7258-7772-0 (PDF)
https://doi.org/10.3390/books978-3-7258-7772-0

Contents

About the Editors

Jun Hu

Jun Hu is a professor at the School of Civil Engineering and Architecture, Hainan University, Haikou, China. His research focuses on sustainable architecture, green and low-carbon buildings, infrastructure sustainability, civil engineering, and structural health monitoring. He has contributed to multiple Special Issues in MDPI journals and has extensive experience in promoting environmentally responsible and sustainable engineering solutions.

Guan Chen

Guan Chen is a researcher at the Earth Observatory of Singapore, Nanyang Technological University, Singapore. His research mainly focuses on geotechnical earthquake engineering, particularly earthquake ground motion characterization and modeling, seismic hazard and risk analysis, and performance-based earthquake engineering. He has led numerous research projects and contributed to several MDPI Special Issues, providing expertise in earthquake engineering reliability and risk assessment.

Yong Fu

Yong Fu is a researcher at the Department of Ocean Science and Engineering, Southern University of Science and Technology, Shenzhen, China. His research interests include marine geotechnical engineering, offshore foundations, underground space technology, and physical experiments and finite element modeling. He has been involved in multiple MDPI Special Issues and has extensive experience in offshore and marine civil engineering research.

Preface

This reprint, *Marine Geotechnical Engineering and Marine Civil Engineering Construction*, presents a collection of recent studies addressing key challenges and emerging developments in offshore and marine civil engineering. The contributions cover a wide range of topics, including offshore foundations, pipeline–soil interactions, cyclic behavior of marine soils, geotechnical improvement technologies, offshore pile foundations, and sustainable marine infrastructure systems.

This volume begins with investigations into offshore pile foundations and lateral bearing mechanisms. Wang et al. study a novel offshore rubber airbag branch pile system and examine its horizontal bearing behavior through numerical simulations and theoretical model development. Their work highlights the role of pile geometry, buried depth, and vacuum-induced resistance mechanisms in improving lateral performance under marine loading conditions. Cheng et al. investigate the uplift behavior of offshore pipelines buried in a spatially varying clayey seabed. By incorporating random field modeling and probabilistic finite element analysis, the study evaluates how spatial variability in soil strength influences uplift resistance and failure probability, providing insights for reliability-based offshore pipeline design.

Several contributions focus on the mechanical behavior and improvement of offshore geomaterials. Liu et al. examine the influence of the mixing degree between coarse and fine particles on the strength characteristics of offshore and coastal foundation materials. Du et al. investigate the stiffness degradation of organic matter-disseminated sand under cyclic loading, providing experimental observations relevant to long-term marine foundation performance. Zhou et al. report an experimental study on the preparation of highly active bacterial suspensions for microbially induced carbonate precipitation (MICP), contributing to sustainable ground improvement technologies for marine environments.

Offshore foundation systems and pile behavior are another major focus of this volume. Wang et al. conduct sensitivity analyses on factors affecting the bearing capacity of suction bucket foundations in soft clay. Guo et al. study the load-carrying behavior of large open-ended pipe piles in cohesionless soils, while Wang et al. develop a new p–y curve formulation for laterally loaded large-diameter monopiles in soft clays. Xu et al. further investigate uplift capacity calculations for suction caissons in sand under different drainage conditions. These studies collectively advance the understanding of offshore foundation responses under complex marine loading and soil conditions.

In addition, Liu et al. investigate geopolymer grouting materials with varied mix proportions, contributing to the development of environmentally friendly and high-performance grouting technologies for marine and underground engineering applications.

Overall, this reprint highlights recent advances in marine geotechnical engineering through a combination of experimental studies, theoretical developments, numerical simulations, and engineering applications. The Guest Editors hope that this collection will serve as a useful reference for researchers, engineers, and practitioners working in offshore engineering, marine infrastructure, and sustainable geotechnical systems, while also encouraging interdisciplinary collaboration and future innovation in the field.

This research was funded by the National Natural Science Foundation of China (No. 51968019, 42467019, 52471281), the Self-Directed Research Fund of the State Key Laboratory of Tropic Ocean Engineering Materials and Materials Evaluation (No. STOEM99272619), the High Technology Direction Project of the Key Research and Development Science and Technology of Hainan Province, China (No. ZDYF2024GXJS001), and the Hainan Provincial Natural Science Foundation Enterprise Talent Project (No. 525QY918).

Jun Hu, Guan Chen, and Yong Fu
Guest Editors

sustainability

MDPI

Article

Study on the Horizontal Bearing Characteristic of a New Type of Offshore Rubber Airbag Branch Pile

Xiaolei Wang [1,2], Zeyuan Wang [1,*], Changfeng Yuan [3] and Libo Liu [1]

1 College of Civil Engineering, Hebei University of Engineering, Handan 056006, China; wangxiaolei@hebeu.edu.cn (X.W.); liulibo@hebeu.edu.cn (L.L.)
2 State Key Laboratory of Coastal and Offshore Engineering, Dalian University of Technology, Dalian 116024, China
3 College of Civil Engineering, Qingdao University of Technology, Qingdao 266033, China; yuanchangfeng@qut.edu.cn
* Correspondence: wzy18832048378@163.com; Tel.: +86-18832048378

Abstract: In this work, a new type of marine rubber airbag branch pile has been presented, and the influences of the exposed length L_0 of the pile, size of rubber airbag branch and depth S_0 of rubber airbag branch embedded in soil on the horizontal bearing capacity of the pile, have been investigated using numerical simulations. Simulation results were used to modify the eigenvalue equations of the horizontal bearing capacity. The results also showed that reverse displacement and the bending moment of the rubber airbag branch pile were lower in pipe piles with larger diameters, and the horizontal bearing capacity was more stable. At small horizontal displacements of the pile top, horizontal bearing capacities of large-diameter pipe piles were slightly higher, while for the pile's top horizontal displacements of above 10 mm, horizontal bearing capacities of rubber airbag branch piles became significantly greater than those of the large-diameter pipe piles. Based on assumptions, the calculation equations of vacuum negative pressure and friction force between the rubber airbag branch and soil were derived. The equations for calculating the characteristic horizontal bearing capacities of rubber airbag branch piles were also derived and were modified based on simulation results. The calculation results confirmed the improvement in the accuracy of the modified equations.

Keywords: rubber airbag branch pile; numerical simulation; horizontal bearing capacity; vacuum negative pressure; equation correction

1. Introduction

With the continuous development and exhaustion of marine resources, particularly the invention of offshore wind power generators as a novel energy source, there is little harmful impact on the environment and is economical and practical; therefore, wind power generators have become extremely popular around the world. Today, about two-thirds of single pile foundations are being used to resist the lateral loads created by sea winds, currents and waves as well as seismic waves. The impact forces of these loads have great impacts on pile horizontal displacements [1]. Based on the analytical results obtained for the horizontal bearing capacities of marine high pile foundations, pile exposed lengths and pile foundation forms were found to have great influences on the horizontal bearing capacities of these structures. In addition, soil resistance around piles varied nonlinearly with depth. Obtaining an accurate calculation of the mechanical characteristics of pile-soil is difficult [2,3]. Therefore, in this research, the form of offshore pile foundations was improved to change their horizontal force transfer mechanisms, which provided a solution for the problem of insufficient horizontal bearing capacity and a large displacement of high pile foundations.

Currently, theoretical models, in situ test methods and finite element numerical models are often used for the investigation of horizontal bearing characteristics of pile foundations.

However, finite element numerical simulations are widely applied because they are not restricted by the environmental conditions of the field [4,5]. Jeong [6] investigated the load distribution and deformation of drilling wellbores under lateral loads through field load tests and P-Y curves, and found that the main factor affecting drilling precision was wellbore stiffness. Mahdi Taha K et al. [7] In order to determine the advantages of the pile with wings, four embedded ratios were used to model the flexible and rigid pile types with different wings and sizes. The results showed that adding wing plates to the pile can improve the lateral force of the pile and greatly reduce the lateral deflection of the pile. Poorjafar Aysan et al. [8] summarize the small-scale laboratory simulation results of pile behavior under a lateral load, considering parameters, such as a short or long pile, a single or group pile, spacing and stiffness or flexibility. The results showed that the short pile had rigid movement, the displacement of the surrounding soil occurred along the total length of the pile, and the pile rotated around a point; however, the long pile had flexible movement in the long part of the pile. Francisco González et al. [9] proposed an equivalent linear boundary element–finite element model that considered soil degradation at the pile–soil interface, which was used to approximate the lateral time-harmonic response of piles. Amar Bouzid Djillali [10] proposed a new P-Y formula containing a new P-Y shape and a new initial stiffness and coded it in the existing Winkler computer program. Through the study of three cases in the literature, the results of the new formula were verified. Deendayal Rathod et al. [11] carried out a series of indoor model tests on instrumented model piles embedded in slopes with different length–diameter ratios. Endi et al. [12] found that the P-Y curves obtained, based on the API standards, were safe, while the "M" method was unsafe; however, the results obtained from the "M" method were closer to the experimentally measured values. Ahmed Sahli et al. [13] discuss the compilation of the calculation program for the safety modeling and verification of the pile limit state in the current structural foundation. Accordingly, the reliability technology is applied to the probability analysis of piles, where the finite element method (FEM) is combined with the boundary element method (BEM), and Mindlin's basic solution is used to model the soil by boundary element, which is suitable for considering three-dimensional infinite half-space.

The cross-section shape of the pile, theoretical analysis and calculation method of pile–soil interactions all have a great influence on the design bearing capacity of the pile. At present, there are few types of research on reducing the horizontal displacement of the pile by applying transverse binding force, and the method of changing the cross-section shape of the pile cannot weaken the relative stiffness of the pile and soil. Although the variable section pile can improve the horizontal bearing capacity of the pile, it is easy to cause the pile-soil separation phenomenon at the variable section of the pile under a long-term cyclic load. Moreover, the study shows that the P-Y curve is more conservative than the M method, and the M method is closer to reality. Therefore, in the current work, a new type of rubber capsule supporting piles was proposed. The ABAQUS software was used to analyze the horizontal bearing characteristics of different rubber airbag branch pile sizes, buried depths S_0, exposed lengths of pile above the soil surface L_0 and horizontal displacements of pile top. The results were compared with those obtained for large-diameter pipe piles. On this basis, the equations for calculating the characteristic values of the horizontal bearing capacity of rubber airbag branch piles were derived.

2. Horizontal Bearing Mechanism of Rubber Air Bag Branch Piles

Rubber airbag branch piles are mainly composed of large-diameter pipe piles and hollow rubber airbag branches. In Figure 1, R is the external radius of the rubber sac ascot and r is the pile diameter. Due to the characteristics of rubber materials, such as high density, low rigidity and easy deformation [14,15] and the physical and mechanical characteristics of marine muddy soft soils with low permeability, the rubber airbag branch compresses soil when it is pressurized in the soil. The air and water in the joint between the surface of the rubber airbag branch and soil are gradually eliminated, and the air pressure is decreased, resulting in negative vacuum pressure. When the pile body is subjected to a horizontal load, the soil and rubber airbag branch work cooperatively and a mutual suction-and-pull action occurs, resulting in the generation of resistance to the horizontal load and an increase in the horizontal bearing capacity of piles. In addition, the cross-sectional areas of rubber airbag branches are larger than those of pile bodies. A relatively closed space is created from the pile end to the rubber airbag branch. When pile bodies undergo horizontal displacement, a vacuum negative pressure is generated on the pile side opposite to the direction of displacement to resist the displacement of the pile, which acts on the pile body in the form of an arc transverse surface load. Because of the circular shape of the rubber airbag branch, no matter whether the pile top is subjected to the load from any horizontal direction, the rubber airbag branch generates lateral restraint on the pile body.

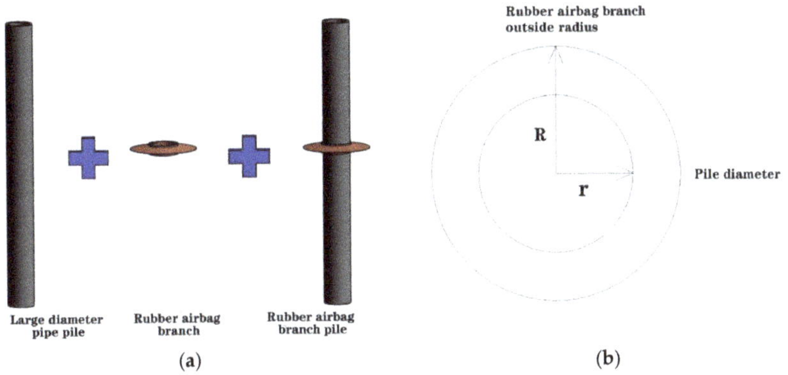

Figure 1. (**a**) Schematic diagram of rubber airbag branch pile; (**b**) plan of rubber airbag branch.

Due to the favorable elastic deformation properties of rubber airbag branch materials, it is difficult for the pile to fully play its role in very small pile displacements. Therefore, due to bearing demand, fine steel wires could be embedded to improve the rubber airbag branch tensile's bearing capacity to decrease large tensile deformations due to the elastic properties of rubber materials during tension.

3. Finite Element Numerical Simulation and Result Analysis

3.1. Establishment of Finite Element Model of Rubber Airbag Branch Pile

The ABAQUS finite element simulation software was applied to establish a rock-socketed pile foundation. The established pile was a large-diameter pipe pile with outer and inner diameters of 5 and 3.25 m, respectively, a wall thickness of 1.75 m and a buried depth of 55 m. Moreover, the rock-socketed depth was 5 m and the pile body exposed lengths L_0 were 0, 5, 10, 15 and 20 m. To decrease the amount of stress transferred from the pile to the soil boundary through a horizontal load, a cylindrical model with a radius of 25 m and depth of 70 m, where 0–50 m was a clay layer and 50–70 m was a rock layer, was adopted. The Mohr–Coulomb constitutive and linear elastic constitutive models were used for the simulation of the clay layer. A linear elastic constitutive model was applied for rock stratum, pile and rubber airbag branch. In order to avoid relative slippage, the pile-rubber airbag branch and pile bottom-rock layer contacts had to be bound. A penalty function was applied for the rubber airbag branch surface-soil body and pile body-soil body contacts. Pile-rock surface contact was considered to be a rough contact. To realize the transformation of pile top horizontal displacement from a small horizontal displacement to a large horizontal displacement, the horizontal displacement range of the pile top was considered as 0~50 mm with an interval of 2 mm. Table 1 summarizes the physical and mechanical parameters of the soil layer, rubber airbag branch and pile, and Figure 2 shows the mesh division of the model.

Table 1. Physical parameters of each component [16].

	Modulus	Poisson Ration	Cohesion	Severe	Internal Friction Angle
Clay layer	30 MPa	0.3	25 kPa	19.23 kN/m	14°
Rock stratum	400 GPa	0.3	——	27 kN/m	——
Rubber airbag branch	10 MPa	0.3	——	10 kN/m	——
Pile	38 GPa	0.3	——	25 kN/m	——

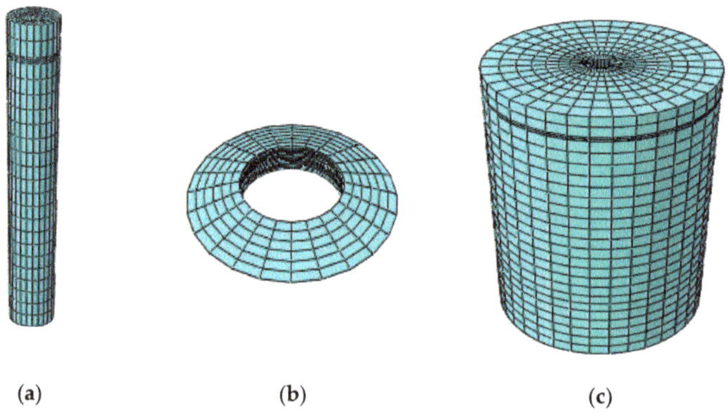

Figure 2. (**a**) Pile grid division; (**b**) rubber airbag branch grid division; (**c**) component grid division.

3.2. Simulation Conditions

By analyzing the horizontal loading mechanism of rubber airbag branch piles, it was concluded that their horizontal bearing capacities were mainly affected by the pile body exposed length L_0, rubber airbag branch size and buried depth S_0 in soil. In this section, these three influencing factors are compared and analyzed under the following combination of working conditions.

The exposed length of high pile foundations is generally long. Without an external force constraint, pile top displacement, due to a horizontal load, is relatively large and bending deformations can easily occur. By comparing rubber airbag branch piles and large-diameter pipe piles at various exposed lengths L_0, the effect of L_0 on horizontal bearing capacities was explored when horizontal displacement deformation occurred on the pile top. Table 2 summarizes the specific combination of working conditions for this test.

Table 2. Combination of different lengths of exposed pile.

R (m)	Height (m)	S_0 (m)	L_0 (m)
5	2	5	0/5/10/15/20

It was found that an increase in the height and outer radius of rubber airbags increased soil contact areas, which affected the negative vacuum pressure and friction force of rubber airbag branches. In addition, with the increase in the rubber airbag branch height, the soil around the pile was separated, affecting the pile's horizontal bearing capacity. In order to evaluate the effect of the rubber airbag branch size on pile top horizontal bearing capacities, both L_0 and S_0 were fixed at 5 m. Tables 3 and 4 summarize the combined working conditions of the test.

Table 3. Combination of different heights.

R (m)	Height (m)	S_0 (m)	L_0 (m)
5	1/1.5/2	5	5

Table 4. Combination of different rubber airbag branch outer radius conditions.

R (m)	Heigh (m)	S_0 (m)	L_0 (m)
3.125/3.75/5	2	5	5

The buried depth S_0 of rubber airbag branch in the soil also affected pile horizontal bearing capacity; especially, horizontal reactions of rubber airbag branch at various pile body positions affected pile bending performance. In order to explore the effects of rubber airbag branch on the two properties of the buried depth S_0 of rubber airbag branch and the length L_0 of pile above the soil surface, the relationship between pile top horizontal displacement and bearing capacity, and that between pile top horizontal displacement and pile body bending moment were compared and analyzed. Table 5 summarizes the combined conditions applied in this experiment.

Table 5. Combination of different outer radius conditions.

R (m)	Height (m)	S_0 (m)	L_0 (m)
5	2	5/10/20	5

3.3. *Result Analysis*

3.3.1. Comparative Analysis of Different Exposed Pile Lengths

According to the working conditions given in Table 2, at the pile top horizontal displacements of 2 mm~50 mm, the comparison results of the exposed lengths L_0 of the straight pile and rubber airbag branch pile outside the pile were different, as shown in Figure 3. It was seen that an increase in L_0 decreased the pile horizontal bearing capacity. The pile top horizontal bearing capacity was increased with the increase in pile top horizontal displacement at a growth rate of an almost straight line; also, the growth rate of the horizontal bearing capacity of rubber airbag branch piles was greater than that of the large-diameter pipe piles. However, the horizontal bearing capacity of rubber airbag branch piles with small pile top horizontal displacements was smaller than that of the large-diameter pipe piles. In Figure 3a, at a pile top horizontal displacement of 12 mm, the horizontal bearing capacities of rubber airbag branches and straight piles were 2827 and 2880 kN, respectively. At a pile top horizontal displacement of 14 mm, the horizontal bearing capacities of rubber airbag branches and vertical piles were 3369 and 3364 kN, respectively. It was seen that the horizontal bearing capacity of the rubber airbag branch pile was higher than that of the large-diameter pipe piles at pile top displacements of 12–14 mm. In Figure 3b–e, the pile top horizontal displacements were 4~6, 4~6, 6~8 and 8~10 mm, respectively. The bearing capacities of the rubber airbag branch piles exceeded those of vertical piles. At L_0 values of 0, 5, 10, 15 and 20 m, the horizontal displacements of rubber airbag branches and large-diameter pipe piles reached 50 mm and the difference in the horizontal loads they were able to carry were 1510, 1891, 1090, 656 and 410 kN, respectively. The obtained results showed that by increasing L_0, the horizontal bearing capacity of the rubber airbag branch pile was first increased and then decreased under large pile top displacements.

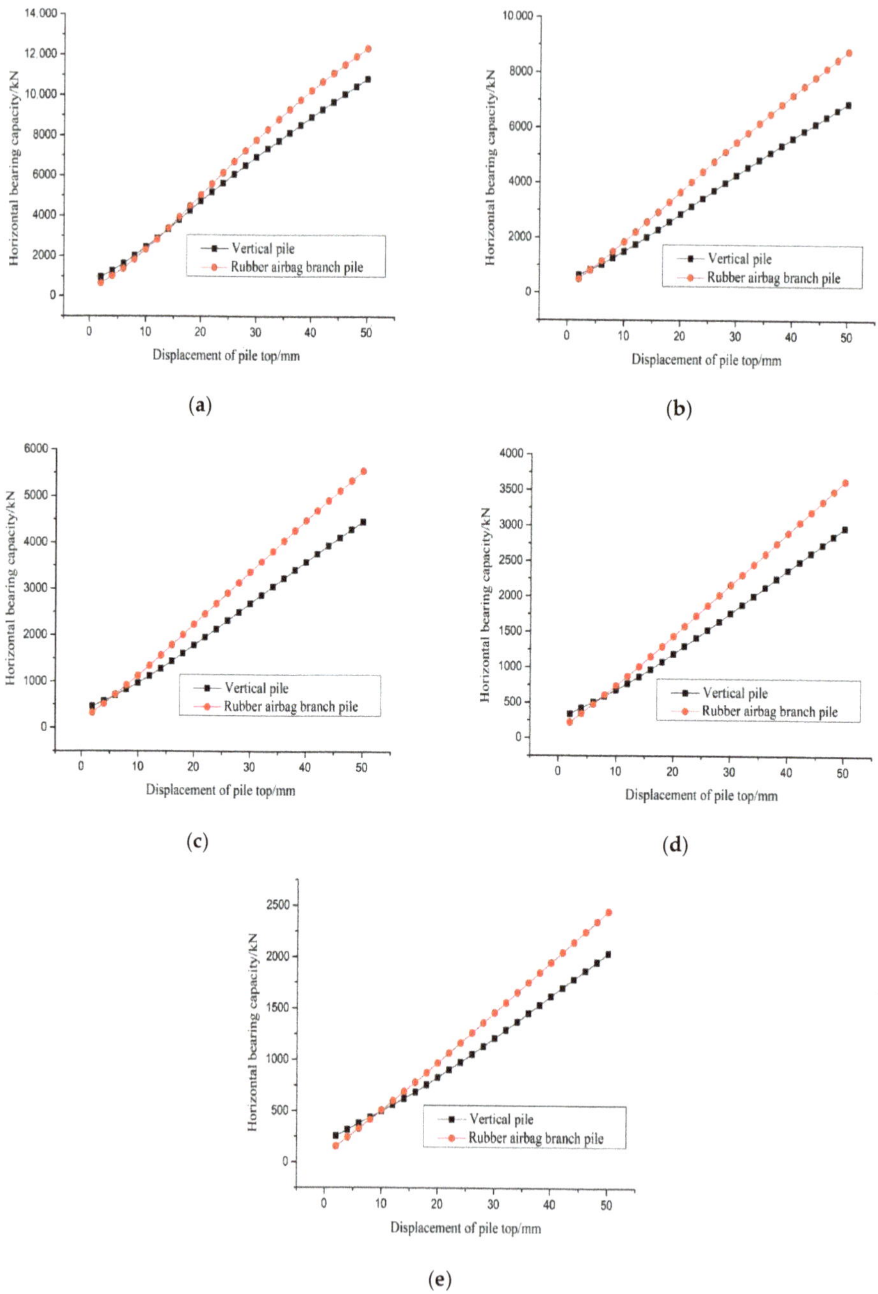

Figure 3. (**a**) 0 m external exposure of pile L_0; (**b**) 5 m external exposure of pile L_0; (**c**) 10 m external exposure of pile L_0; (**d**) 15 m external exposure of pile L_0; (**e**) 20 m external exposure of pile L_0.

Based on the five comparison results mentioned above, the effect of L_0 on horizontal displacement was generated from the fact that when the pile top was exposed to horizontal loads, the flexural rigidity of the pile first played a role and then transferred to the soil under the constraint of the soil reaction around the pile. However, the length L_0 of the pile above the soil surface directly depended on flexural strength, which resulted in a decreased horizontal bearing capacity of the pile. As the length of the pile's exposed part was increased, the distance of horizontal load transfer to soil was increased and the extension of time for the reaction of soil around the pile to exert its effect along with the horizontal bearing capacity of the pile was also decreased. When the rubber airbag branch entered the soil along with the pile, some parts of the soil and pile were separated. In the process of pile top horizontal load transfer, part of the soil reaction around the pile was missing. Therefore, when the pile top horizontal load was transferred to the rubber airbag branch, the rubber airbag branch first produced tensile deformation with pile body displacement direction and, then, extrusion deformation occurred in the opposite direction of the pile body horizontal displacement. In this process, the horizontal load-bearing capacity of rubber airbag branch piles was not high enough to make up for the loss of soil reaction around the pile and the horizontal bearing capacity of rubber airbag branch piles was less than those of large-diameter pipe piles. However, as pile top horizontal displacement occurred, the tensile and extrusion deformations of rubber airbag branches gradually reached their limit values. Pile horizontal bearing capacity was increased due to the horizontal load resistance generated by rubber airbag branches.

3.3.2. Comparative Analysis of Different Sizes of Rubber Airbag Branches

By comparing the working conditions given in Tables 3 and 4, Figures 4–6 were obtained. It was seen from Figure 4 that, the horizontal bearing capacities of rubber airbag branch piles with different heights were greater than those of large-diameter pipe piles and the increased rates for horizontal bearing capacities with pile top horizontal displacements were greater than those of straight piles. At a pile top horizontal displacement of 50 mm, the horizontal bearing capacity of the straight pile was 6876 kN, while the horizontal bearing capacities of the rubber airbag branch pile were 8923, 9039 and 8767 kN at disc heights of 1, 1.5 and 2 m, respectively. It was found that the bearing capacity of the rubber airbag branch pile was 2163 kN higher than that of the straight pile at a pile height of 1.5 m. Based on the findings of quantitative analysis, the rubber airbag branch height was 1 m, which was greater than the horizontal bearing capacity at pile heights of 2 m and less than 1.5 m. It was seen that too large separation areas of pile and soil negatively affected the function of rubber airbag branches. An increase in rubber airbag branch height did not always improve pile horizontal bearing capacity.

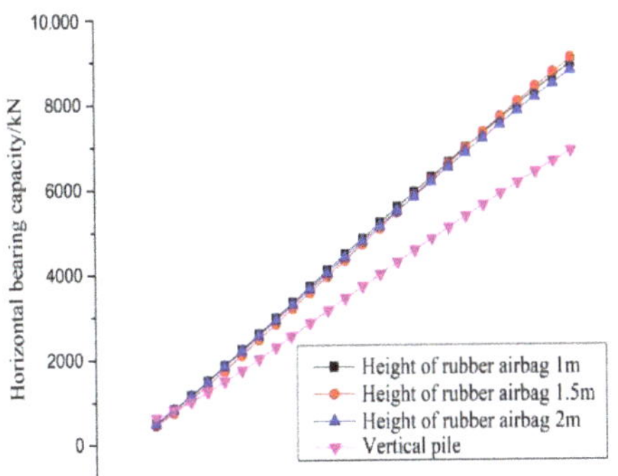

Figure 4. Influence of rubber airbag branch height on horizontal bearing capacity of pile.

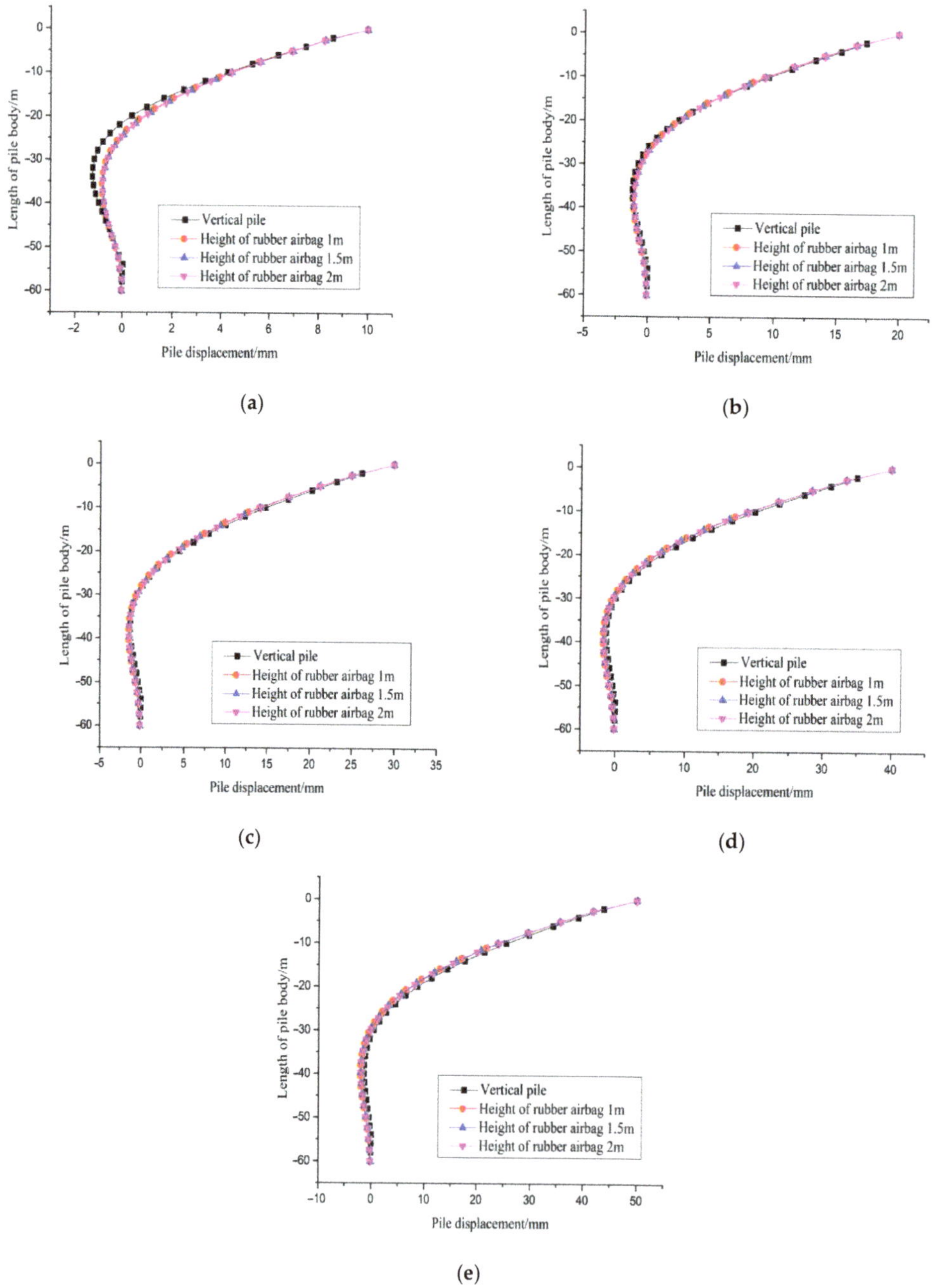

Figure 5. (**a**) 10 mm pile top horizontal displacement; (**b**) 20 mm pile top horizontal displacement; (**c**) 30 mm pile top horizontal displacement; (**d**) 40 mm pile top horizontal displacement; (**e**) 50 mm pile top horizontal displacement.

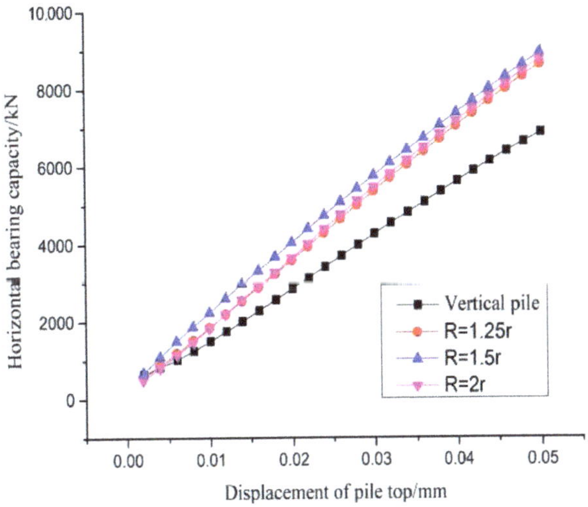

Figure 6. The influence of the outer radius of rubber airbag branch on horizontal bearing capacity.

To understand the effect of rubber airbag branch height on pile body horizontal displacement, pile body horizontal displacements of 10, 20, 30, 40 and 50 mm were analyzed. As shown in Figure 5, compared with the large-diameter pipe piles, rubber airbag branches imposed certain constraints on pile horizontal displacements. However, increasing rubber airbag branch height had little effect on pile horizontal displacement. In Figure 5a, the reverse horizontal displacement position along the pile body showed that the reverse horizontal displacement position of the large-diameter pipe pile was 26.2 m, while at the rubber airbag branch heights of 1, 1.5 and 2 m, the reverse horizontal displacement positions of the rubber airbag branch piles were 26.5, 26.8 and 27.5 mm, respectively. It was seen that the reverse horizontal displacement point of the rubber airbag branch pile was lower than that of the large-diameter pipe pile and the reverse horizontal displacement point of the rubber airbag branch pile was gradually decreased with an increase in the rubber airbag branch height.

It can be seen in Figure 6 that pile horizontal bearing capacity could be improved by increasing the outer radius R of the rubber airbag branch. The horizontal bearing capacities of the piles were 8641, 8925 and 8767 kN for a pile top displacement of 50 mm, rubber airbag branch height of 2 m and the R values were 1.25 and 1.5 and $2r$, respectively. Compared with large-diameter pipe piles, horizontal bearing capacities were increased by 1765, 2049 and 1891 kN, respectively, and a maximum horizontal bearing capacity was obtained for an r value of $1.5R$. An increase in horizontal bearing capacity showed that pile horizontal bearing capacity could not be continuously increased by the increase in the rubber airbag branch outer radius.

3.3.3. Comparative Analysis of Different Rubber Airbag Branch Buried Depths S_0

Figures 7 and 8 were derived by comparing the different working conditions of rubber airbag branch buried depth S_0 values. It can be seen in Figure 7 that the maximum pile bearing capacities of 8767, 8502 and 8343 kN were obtained for the S_0 values of 5, 10 and 20 m, respectively. It was concluded that deeper rubber airbag branches decreased pile bearing capacities. When pile top horizontal displacement was within 4 mm, straight pile bearing capacity was greater than that of the rubber airbag branch. Pile top horizontal displacement was within 18 mm and pile bearing capacity was smaller than that of the pile at a buried depth of rubber airbag branch at 5 m. It was seen that the soil reaction loss around the pile, due to different buried depths of rubber airbag branches, affected the improvement of pile horizontal bearing capacity.

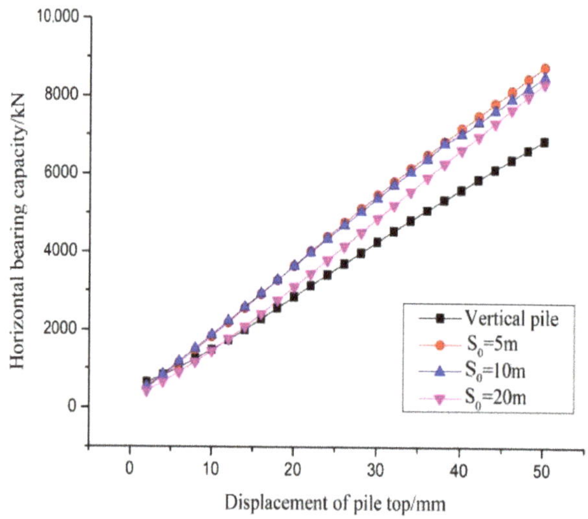

Figure 7. Influence of buried depth S_0 of rubber airbag branch on bearing capacity.

By evaluating the effects of different buried depths of rubber airbag branches on pile bending moments, as shown in Figure 8, it was clearly seen that the bending moment of large-diameter pipe piles in the first and second inflection points appeared near the pile body of 5 and 8 m, respectively. However, the bending moment was changed in a relatively gentle trend after the first inflection point at 5 m of the rubber airbag branch pile. Analyses showed that the reason for the second turning point of a large-diameter pipe pile was that when the soil reaction around the pile increased with the depth, the rubber airbag branch pile and soil reaction around the pile resisted the pile top horizontal load under the action of the rubber airbag branch. Therefore, there were no obvious inflection points near the pile body of 8 m. From Figure 8a, it was concluded that the maximum bending moment of the rubber airbag branch was less than that of a large-diameter pipe pile at a pile top horizontal displacement of 10 mm. The maximum bending moment was slowly decreased with the increase in S_0. Compared with Figure 8, the maximum bending moment of rubber airbag branch piles was lower than those of large-diameter pipe piles and the maximum bending moment was increased by increasing the pile top horizontal displacement. At a pile top horizontal displacement of 30 mm and with S_0 values of 5 and 10 m, the maximum pile body bending moments were greater than those of large-diameter pipe piles. At a pile top horizontal displacement of 50 mm and with S_0 values of 5 to 20 m, the maximum bending moments of the rubber airbag branch pile bodies exceeded those of the large-diameter pipe piles. However, from the overall curve change in the pile bending moment, the variation range of the rubber airbag branch pile curve is smaller, and the horizontal bearing performance of the rubber airbag branch pile is more stable than that of large-diameter pipe piles.

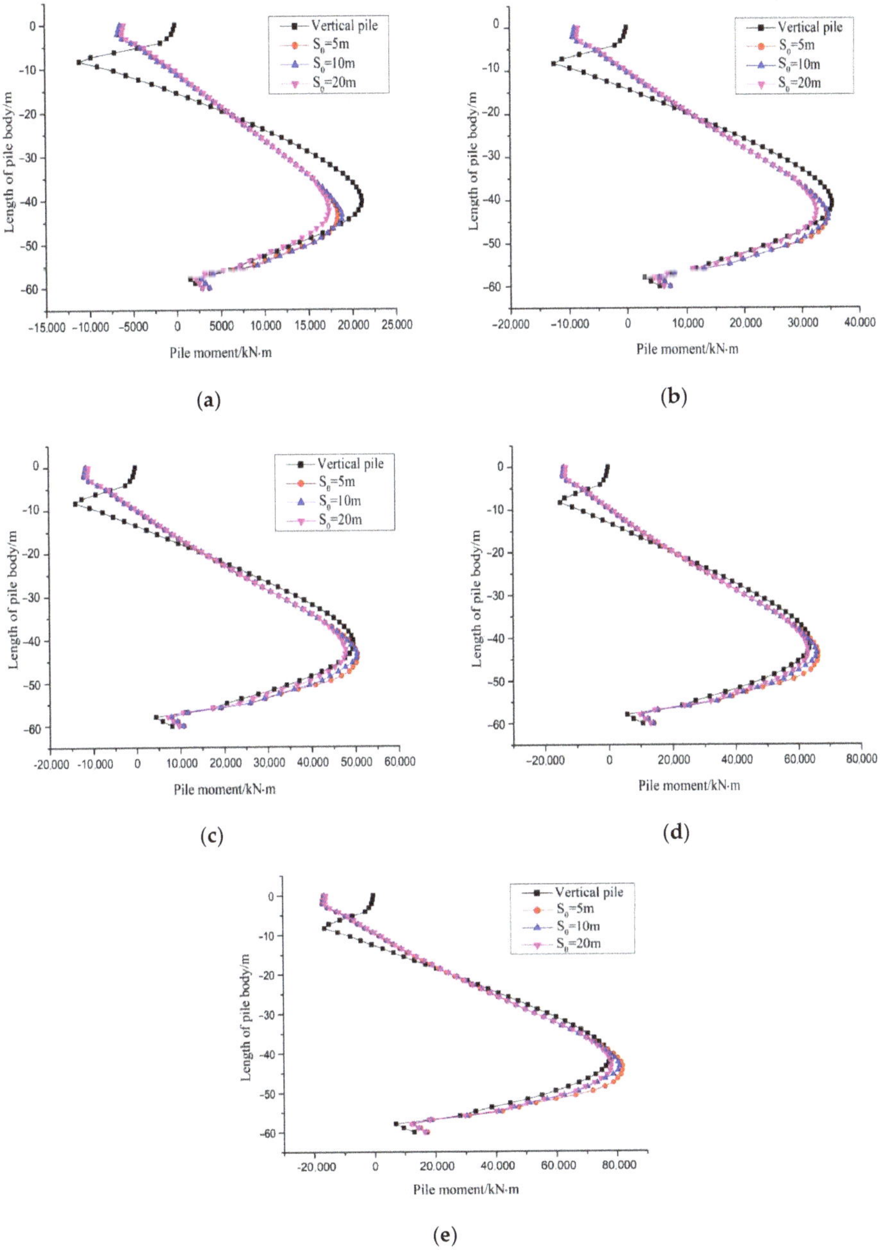

Figure 8. (**a**) 10 mm horizontal pile top displacement; (**b**) 20 mm horizontal pile top displacement; (**c**) 30 mm horizontal pile top displacement; (**d**) 40 mm horizontal pile top displacement; (**e**) 50 mm horizontal pile top displacement.

4. The Horizontal Bearing Capacity Characteristic Value of Rubber Airbag Branch Piles

4.1. Formula Correction for Bearing Capacity Characteristic Value of Large Diameter Tubular Piles

When there were no measured horizontal bearing capacities for single piles, the characteristic value of pile horizontal bearing capacity played an important role in engineering and it was still taken as important reference data in designs [17]. The equations of the horizontal bearing capacity characteristic value of a single pile are given in China's existing standards. Using these equations, the horizontal bearing capacity of the rubber airbag branch pile was studied. The equation of the horizontal bearing capacity characteristic value is given in Equation (1):

$$R_{ha} = 0.75\frac{a^3 EI}{V_x}X_{0a},\ a = \sqrt[5]{\frac{EI}{mb_0}} \tag{1}$$

where R_{ha} is the horizontal bearing capacity characteristic value; "a" is the pile horizontal deformation coefficient; EI is pile bending stiffness; X_{0a} is the allowable pile top horizontal displacement; V_x is the pile top horizontal displacement coefficient; m is the counterforce growth coefficient of horizontal foundation; b_0 is the pile calculation width.

The above equation does not take into account pile exposed length. Generally, this equation can be applied for piles with short exposed lengths and pile diameters of less than 0.8 m, while large-diameter piles are employed for marine pile foundations [18,19]. Therefore, the applicability of the equation was verified by comparing the results obtained from the normative equation with numerical simulation results. The horizontal bearing capacities of small-diameter pipe piles with a pile length of 25 m, an outer diameter of 0.8 m and an inner diameter of 0.52 m were compared with the above-mentioned large-diameter pipe pile, as shown in Figures 9 and 10. After analyzing relevant specifications, the value of "M" in the equation was taken as 600 MN/m^4. After calculation, it can be seen in the figure that the normative equation was more suitable for pipe piles with short exposed lengths and small pile diameters, but not for large-diameter pipe piles.

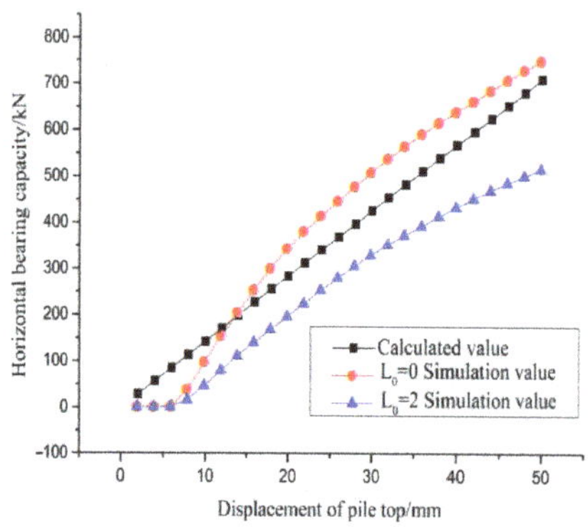

Figure 9. Comparison of horizontal bearing capacities of small-diameter pipe piles.

Because of the limitations of the applicable scope of the normative equation, the exposed length L_0 of a large-diameter pipe pile was taken as the most influential parameter. A correction factor was added to the canonical equation, as expressed in Equation (2):

$$\zeta_1 = tL_0^3 + uL_0^2 + vL_0 + w \tag{2}$$

where L_0 is the pile exposed length; t, u, v and w are undetermined coefficients; and ζ_1 is the correction coefficient of the equation for large-diameter piles.

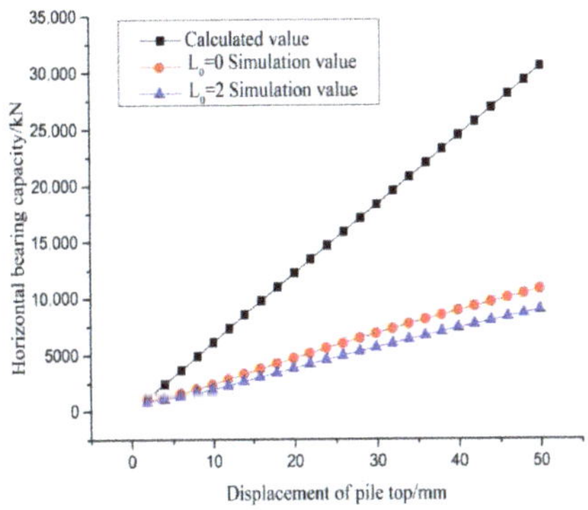

Figure 10. Comparison of horizontal bearing capacities of large-diameter pipe piles.

As shown in Figure 11, the variation in horizontal bearing capacities of large-diameter pipe piles with L_0 values of 0, 5, 10, 15 and 20 m were analyzed in a pile top displacement range of 2–50 mm. It was seen from the figure that pile horizontal bearing capacity was linearly changed with the increase in pile top horizontal displacement. It was also seen from the figure that by increasing pile top horizontal displacement, the horizontal bearing capacity of the pile was changed almost linearly. By analyzing the difference between the calculated and simulated values, the variation law of the horizontal bearing capacity for large-diameter pipe piles with exposed length was derived. As shown in Figure 12, the relationship between the calculated value-simulated value ratio and pile exposed length was fitted; it was also seen that the error after fitting was lower and the values of undetermined coefficients t, u, v and w were obtained. The modified calculation equation of large-diameter pipe pile bearing capacity was stated as Equation (3).

$$R_{\text{ha}} = (-0.02L_0^3 + 1.28L_0^2 - 31.47L_0 + 353.19) \times 10^{-3} \times 0.75\frac{a^3EI}{V_x}x_{0a} \quad (3)$$

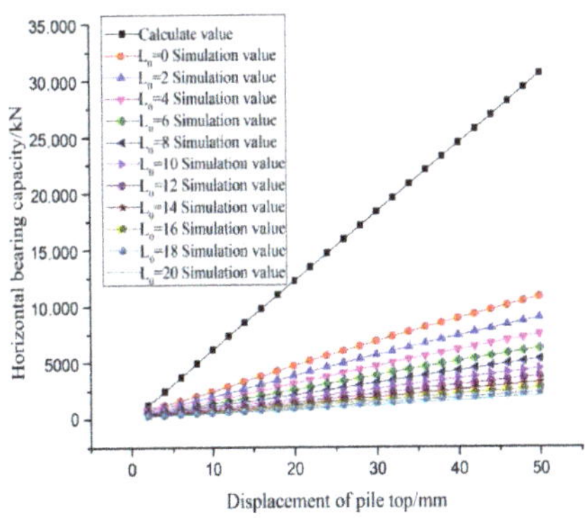

Figure 11. Horizontal bearing capacity of large-diameter pipe piles with different exposed lengths.

Characteristic, simulated and modified normative values of large-diameter pipe piles bearing capacity were compared at L_0 values of 0, 4, 12 and 20 m. As shown in Figure 13, it can be seen that the modified characteristic value of horizontal bearing capacity of large-diameter pipe piles was reasonable.

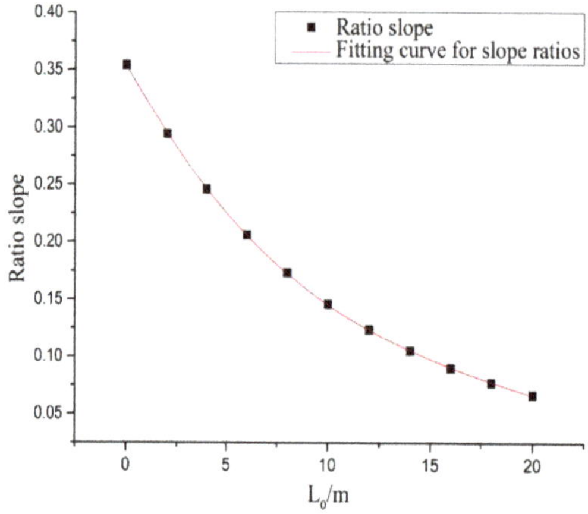

Figure 12. ζ_1 fitting correction coefficients.

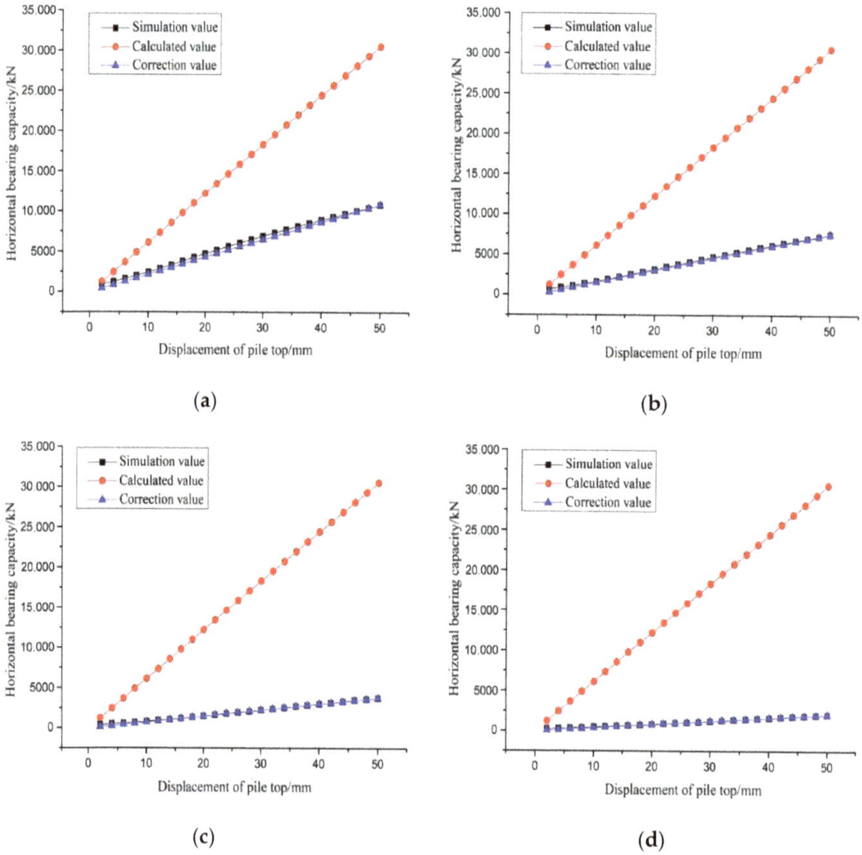

Figure 13. (**a**) 0 m external pile exposure; (**b**) 4 m external pile exposure; (**c**) 12 m external pile exposure; (**d**) 20 m external pile exposure.

4.2. The Characteristic Value of Bearing Capacity for Large-Diameter Rubber Airbag Branch Piles

4.2.1. The Equation of the Resistance of Horizontal Load for Rubber Airbag Branch Piles

Vacuum negative pressure was mainly generated due to soil density as well as a cohesive force and suction force among water molecules. The resistance to a horizontal load due to a rubber airbag branch pile lateral displacement in clay was similar to the processes of drainage and consolidation in low-permeable soils, which required great suction forces to discharge. Refs. [20–22] In this section, negative vacuum pressure and horizontal resistance due to negative friction force were adopted as the main reference forces. In addition, the following assumptions were made: rubber airbag branch pile tension was in the form of small deformation; at a pile top displacement of 0, the rubber airbag branch did not play a role and F was considered to be 0; the soil was uniform and atmospheric pressure inside the soil was not changed by the depth and remained at a constant value. The equations of the horizontal resistance of the rubber airbag branch pile were derived as presented in Equations (4)–(6):

$$F = F_1 + 2F_2 \tag{4}$$

$$F_1 = P\frac{(R+r)h}{2} \tag{5}$$

$$F_2 = \frac{k\sum \gamma h \tan\alpha + c}{\tan\psi} = \frac{k\sum \gamma h \tan\delta + c}{h/(R+r)} \tag{6}$$

where F is the resultant force of negative vacuum pressure and horizontal friction; F_1 is negative vacuum pressure; F_2 is horizontal friction; P is pressure; R is the outer radius of the rubber airbag branch; r is pile radius; h is the rubber airbag branch dish height; ψ is the rubber airbag branch horizontal plane and surface projection angle, $\psi = \arctan[h/(R + r)]$; k is the coefficient of soil side pressure, $k = \upsilon/(1 + \upsilon)$; υ is Poisson's ratio; γ is the soil weight, take the value of 19.23 kN/m^3; δ is the soil internal friction angle, take the value of 14; c is the soil mass cohesive force, take the value of 25 kN/m^2. Other coefficients can be obtained from the text or conversion.

4.2.2. Establishment of the Characteristic Value of Horizontal Bearing Capacity

The rubber airbag branch pile was found to be similar to the variable section pile and could be divided into different sections based on the position of the rubber airbag branch pile installed in the pile. However, rubber airbag branch rigidity was greatly different from pile rigidity and therefore, an equivalent substitution method could not be applied for making the equivalent substitution between the rubber airbag outside radius and pile diameter. Therefore, in this section, it was assumed that the foundation reaction coefficient did not change with the pile depth and deformation at the divided pile section, which could be continuous without additional stress. By assuming the displacement and rotation angle of the rock-socketed pile bottom to be zero, the rubber airbag branch horizontal force acted on the pile body in the form of the resultant force. Employing boundary conditions at the pile bottom, a backward calculation method was developed. A single rubber airbag branch pile was divided into three sections, as shown in Figure 14. Similarly, when the pile body acted on n rubber airbag branches, its amplitude was divided into $2n + 1$ sections. Under the action of horizontal load, the shear force transferred along the pile body and created a horizontal displacement and the pile section at the rubber airbag branch received a reverse horizontal resultant force of F. The equations of single branch and n branches were established as Equations (7) and (8).

$$Q_1 + Q_2 + Q_3 - F = \xi_1\xi_2\xi_3 \times 0.75\frac{a^3EI}{V_x}(\Delta x_1 + \Delta x_2 + \Delta x_3) \tag{7}$$

$$Q_1 + Q_2 + \ldots + Q_{2n+1} - nF- = \xi_1\xi_2\xi_3 \times 0.75\frac{a^3EI}{V_x}(\Delta x_1 + \Delta x_2 + \ldots \Delta x_{2n+1}) \tag{8}$$

where $\sum_{i=0}^{n} Q_{2n+1}$ is the pile horizontal bearing capacity with n rubber airbag branches; $\sum_{i=0}^{n} \Delta x_{2n+1}$ is the pile top horizontal displacement with n rubber airbag branches, (n = 0, 1, 2, 3, …); and ζ_2 and ζ_3 are correction factors related to pile exposed length L_0 and rubber airbag branch buried depth S_0 in the equation of rubber airbag branch pile, respectively. Other parameters are the same as above.

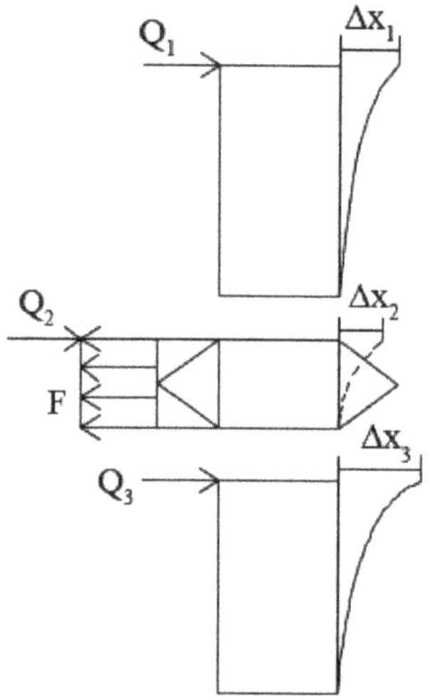

Figure 14. Diagram of force exerted on rubber capsule branch.

The calculation and simulation values of the horizontal bearing capacity of the rubber airbag branch pile are compared. By applying the slope ratio method, as shown in Figures 15 and 16, the ζ_2 and ζ_3 correction coefficients could be obtained, as stated in Equations (9) and (10), respectively.

$$\zeta_2 = (0.112L_0^3 - 4.38L_0^2 + 32.23L_0 + 1058.89)/10^3 \tag{9}$$

$$\zeta_3 = 3.42 \times 10^{-3}S_0 + 1.14 \tag{10}$$

where S_0 is the rubber airbag branch pile's buried depth.

4.3. Verification of Bearing Capacity Eigenvalue Equation of Rubber Airbag Branch Pile

According to the working conditions described in Tables 2 and 3, calculated and simulated horizontal bearing capacity characteristic values before and after modification were compared and analyzed, as shown in Figures 17 and 18. It can be seen from the figure that, at a pile top horizontal displacement of 50 mm and with L_0 values of 0, 5, 10, 15 and 20 m, the differences between corresponding, uncorrected, calculated, and simulated values were 19,100, 22,663, 25,881, 27,802 and 28,978 kN, and those between the corrected, calculated, and simulated values were 1389, 803, 588, 378 and 351 kN, respectively. At S_0 values of 5, 10 and 20 m, the corresponding differences were 22,663, 21,992 and 23,087 kN, respectively. The differences between the modified calculated and simulated values were 803, 936 and 830 kN, respectively. By analyzing the difference values, the corrected calculated values were found to be close to the simulated values and the corrected error was controlled within the range of 10%. Due to the complexity of soil mass' physical and mechanical parameters, there were too many influencing factors, making it impossible to accurately calculate the contact force between the pile–soil and rubber airbag branch pile–soil mass interfaces. Moreover, the correction coefficient of fitting ζ_1, ζ_2, ζ_3

had certain errors. By multiplying corrected coefficients ζ_1, ζ_2 and ζ_3, the overall error was amplified and it was concluded that the horizontal bearing capacity characteristic value of the modified rubber airbag branch pile was greater than the simulated value. However, a comparison of the calculation and comparison results revealed that the calculated values after the addition of the correction coefficients ζ_2 and ζ_3 were closer to the simulation values, and their accuracy was higher than the calculated values before adding correction coefficients, which played a certain role in the accuracy of the calculation results.

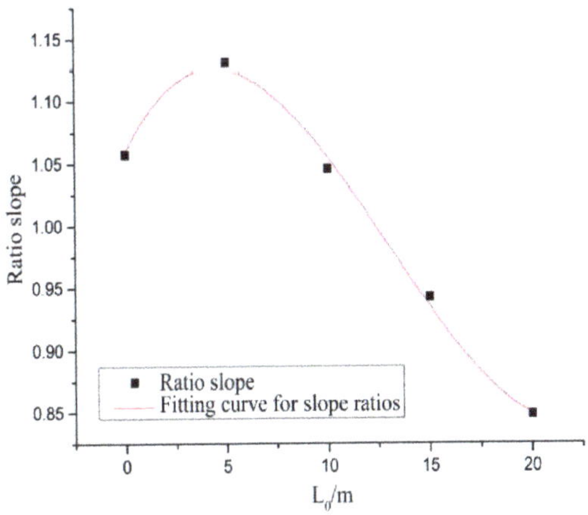

Figure 15. ζ_2 fitting correction coefficient.

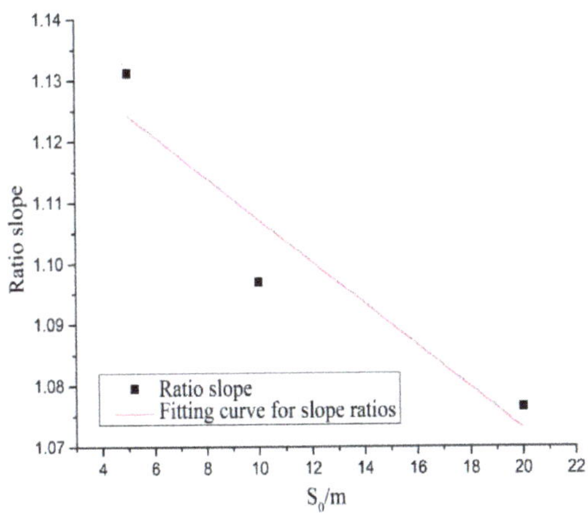

Figure 16. ζ_3 fitting correction coefficient.

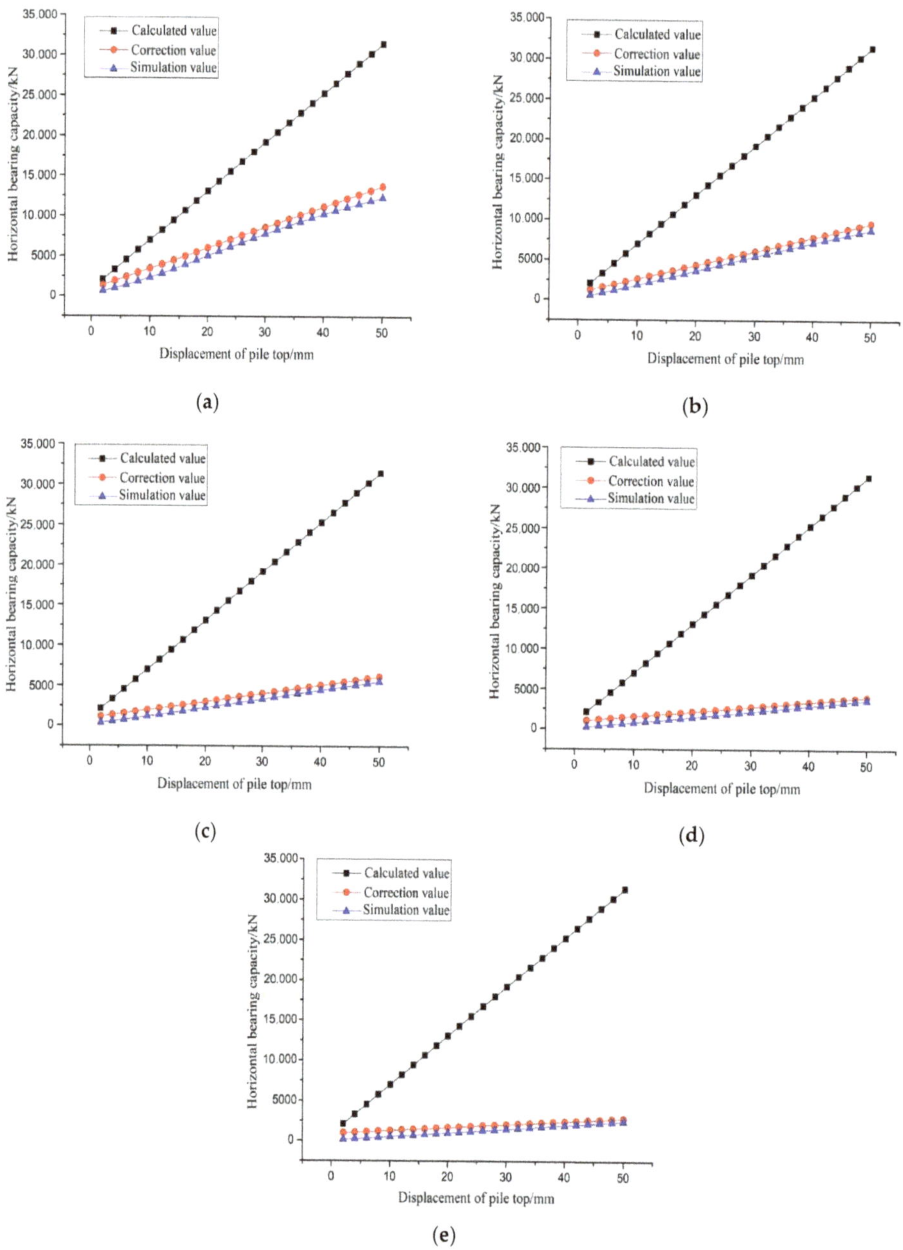

Figure 17. (**a**) 0 m external pile exposure; (**b**) 5 m external pile exposure; (**c**) 10 m external pile exposure; (**d**) 15 m external pile exposure; (**e**) 20 m external pile exposure.

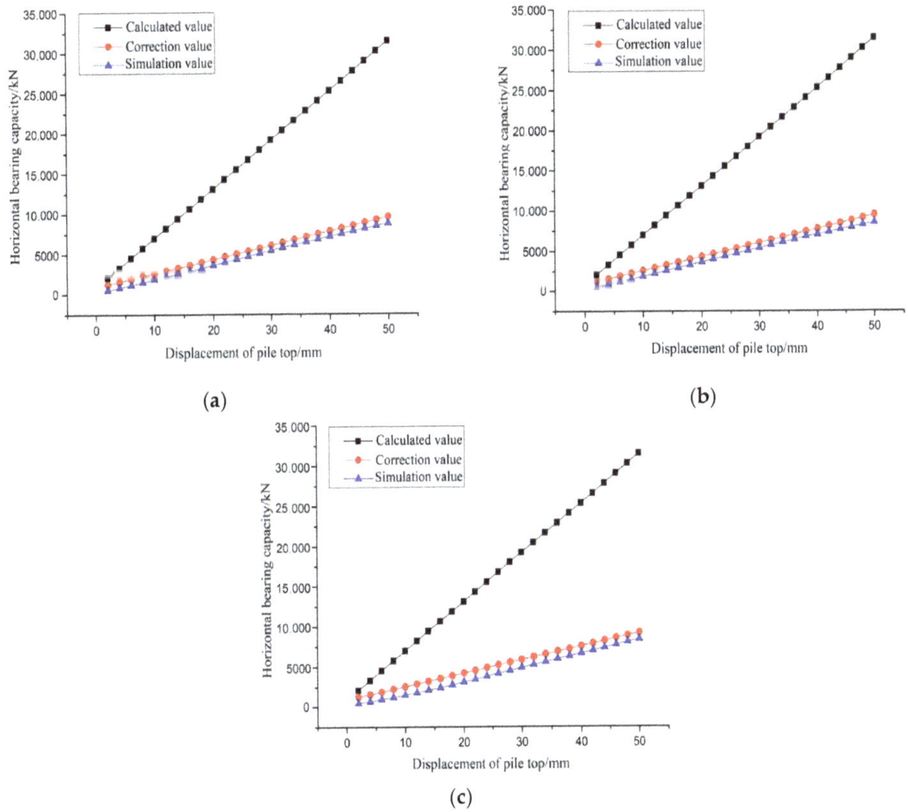

Figure 18. (**a**) 5 m burial depth of rubber airbag branch; (**b**) 10 m burial depth of rubber airbag branch; (**c**) 20 m burial depth of rubber airbag branch.

5. Conclusions

Using the ABAQUS numerical simulation software, a rubber airbag branch pile with a diameter of 5 m was developed to combine the pile exposed length and different pile top horizontal displacements to explore the horizontal bearing performance of piles. The relationship between pile top horizontal displacement and pile horizontal bearing capacity was also investigated under different rubber airbag branch sizes and soil buried depths. The following conclusions were made:

(1) The horizontal bearing capacities of piles with large pile top displacements were obviously higher than those of the pipe piles under the action of a rubber airbag branch bearing tray, which indicated that the rubber airbag branch bearing had a transverse constraint on the pile. As L_0 was increased, pile horizontal bearing capacity was also rapidly decreased.

(2) The increase in the rubber airbag branch height and radius did not increase pile horizontal bearing capacity, however, an increase in pile top horizontal displacement enhanced the increase rate of horizontal bearing capacity compared to that of the large-diameter pipe piles. Under the lateral constraint of the rubber airbag branch pile, the reverse displacement position and reverse pile bending moment were decreased and pile horizontal bearing capacity was more stable.

(3) A buried depth of rubber airbag branch also increased pile horizontal bearing capacity. The horizontal bearing capacity was increased with an increase in the buried depth from 5 to 10 m.

(4) The equation for the rubber airbag branch bearing capacity characteristic value was also derived. After modification, the modified equation of the horizontal bearing capacity characteristic value of the rubber airbag branch pile could be applied as a reference in designs.

Author Contributions: Methodology, X.W.; software, Z.W.; validation, X.W., Z.W. and C.Y.; formal analysis, Z.W.; investigation, Z.W.; resources, C.Y.; data curation, Z.W.; writing—original draft preparation, Z.W.; writing—review and editing, X.W.; visualization, C.Y.; supervision, C.Y.; project administration, L.L. All authors have read and agreed to the published version of the manuscript.

Funding: China Youth Science Foundation Project, 51708317.

Data Availability Statement: The data presented in this study are available on request from the corresponding author. The data are not publicly available due to the follow-up research.

Conflicts of Interest: The authors declare no conflict of interest.

References

1. Doherty, P.; Gavin, K. Laterally loaded monopile design for offshore wind farms. *Proc. ICE-Energy* **2011**, *165*, 7–17. [CrossRef]
2. Kim, Y.; Jeong, S. Analysis of soil resistance on laterally loaded piles based on 3D soil-pilin-teraction. *Comput. Geotech.* **2010**, *38*, 248–257. [CrossRef]
3. Nghiem, H.M.; Chang, N.-Y. Pile under torque in nonlinear soils and soil-pile interfaces. *Soils Found.* **2019**, *59*, 1845–1859. [CrossRef]
4. Wang, H.; Yao, J.; Liu, L. The Application of ANSYS in the Horizontal Bearing Pile. *Appl. Mech. Mater.* **2014**, *193–194*, 935–938. [CrossRef]
5. Edro-Tomislav Simic-Silva, B.M.; Simic, D. 3D simulation for tunnelling effects on existing pile. *Comput. Geotech.* **2020**, *124*, 67–82.
6. Jeong, S.; Kim, Y. Analysis for Laterally Loaded Offshore Piles in Marine Clay. *Geotech. Eng. J. SEAGS AGSSEA* **2020**, *51*, 326–342.
7. Mahdi, T.K.; Al-Neami, M.A.; Rahil, F.H. Experimental and Numerical Study on the Winged Pile-Soil Interaction under Lateral Loads. In *IOP Conference Series: Earth and Environmental Science*; IOP Publishing: Bristol, UK, 2022; Volume 961, pp. 151–165.
8. Aysan, P.; Mahzad, E.F.; Hooshang, K. Pile-soil interaction determined by laterally loaded fixed head pile group. *Geomech. Eng.* **2021**, *26*, 13–25.
9. González, F.; Padrón, L.A.; Aznárez, J.J.; Maeso, O. Equivalent linear model for the lateral dynamic analysis of pile foundations considering pile-soil interface degradation. *Eng. Anal. Bound. Elem.* **2020**, *119*, 59–73. [CrossRef]
10. Djillali, A.B. Analytical Quantification of Ultimate Resistance for Sand Flowing Horizontally around Monopile: New p-y Curve Formulation. *Int. J. Geomech.* **2021**, *21*, 319–335.
11. Rathod, D.; Muthukkumaran, K.; Sitharam, T.G. Effect of Slope on p-y Curves for Laterally Loaded Piles in Soft Clay. *Geotech. Geol. Eng.* **2018**, *36*, 1509–1524. [CrossRef]
12. Zhai, E.; Xu, H.; Guo, S.; Jin, H.; Du, X. Comparative study on horizontal bearing characteristics of offshore wind pipe pile foundation in xiangshu. *Acta Sol. Sin.* **2019**, *40*, 681–686.
13. Sahli, A.; Alouach, M.A.K.; Sahli, S.; Karas, A. Reliability techniques and Coupled BEM/FEM for interaction pile-soil. *Recl. Mécanique* **2017**, *2*, 112–124.
14. Li, Y.M.; Ma, Y.H.; Zhao, G.F.; Zhou, F.L. Experimental Study on the Effect of Alternating Ageing and Sea Corrosion on Laminated Natural Rubber Bearing's Tension-shear Property. *J. Rubber Res.* **2020**, *23*, 151–161. [CrossRef]
15. Cherezova, E.N.; Karaseva Yu, S.; Momzyakova, K.S. Hydrophilic Rubber Based on Butadiene–Nitrile Rubber and Phytogenic Powdered Cellulose. *Polym. Sci. Ser. D* **2022**, *15*, 118–121. [CrossRef]
16. Hu, Y.; Lu, Z.; Zhai, Q.; Zhang, Y. Discussion on p-y curve of large diameter and wing pile in soft clay. *Water Conserv. Hydropower Technol.* **2018**, *49*, 143–152.
17. Zhang, Z.; Guan, P.; Xu, J.; Wang, B.; Li, H.; Dong, Y. Horizontal Loading Performance of Offshore Wind Turbine Pile Foundation Based on DPP-BOTDA. *Appl. Sci.* **2020**, *10*, 492. [CrossRef]
18. Zhai, Q.; Xiang, W.; Li, Y. Study on the Lateral Deformation of the Flexible Berthing Pile of High-Pile Wharf under Ship Impact Load. *Appl. Mech. Mater.* **2015**, *3843*, 1184–1187. [CrossRef]
19. Zhao, M.; Jiang, C.; Cao, W.; Liu, J. Catastrophic model for stability analysis of high pile-column bridge pier. *J. Cent. South Univ. Technol.* **2007**, *14*, 725–729. [CrossRef]
20. Bergado, D.T.; Pitthaya, J.; Pornkasem, J.; Suched, L.; Chartchai, P.; Nuttapong, K.; Francisco, B. Case study and numerical simulation of PVD improved soft Bangkok clay with surcharge and vacuum preloading using a modified air-water separation system. *Geotext. Geomembr.* **2022**, *50*, 137–153. [CrossRef]
21. Almeida, M.S.S.; Deotti, L.O.G.; Almeida, M.C.F.; Marques, M.E.S.; Cardoso, I.M. Vacuum Preloading on Structured Clay: Field, Laboratory, and Numerical Studies. *Int. J. Geomech.* **2021**, *21*, 362–376. [CrossRef]
22. Apriadi, D.; Barnessa, R.A.; Marsa, N.A.I. Finite Element Study of Vacuum Preloading and Prefabricated Vertical Drains Behavior for Soft Soil Improvement. *J. Tek. Sipil* **2019**, *26*, 189. [CrossRef]

sustainability

Article

Uplift Behavior of Pipelines Buried at Various Depths in Spatially Varying Clayey Seabed

Po Cheng [1], Jia Guo [2], Kai Yao [3,4], Chaofan Liu [5], Xiushui Liu [6] and Fei Liu [7,*]

1 State Key Laboratory of Water Resources and Hydropower Engineering Science, Institute of Engineering Risk and Disaster Prevention, Wuhan University, 299 Bayi Road, Wuhan 430072, China; cechengpo@163.com
2 Department of Civil Engineering and Mechanics, Faculty of Engineering, China University of Geoscience (Wuhan), Wuhan 430074, China; jiaguo@cug.edu.cn
3 School of Qilu Transportation, Shandong University, 12550 East Second Ring Road, Jinan 250002, China; yaokai@sdu.edu.cn
4 Shenzhen Research Institute, Shandong University, A301 Virtual University Park, South District of Shenzhen, Shenzhen 518057, China
5 Key Laboratory of Geotechnical and Underground Engineering of Ministry of Education, Department of Geotechnical Engineering, College of Civil Engineering, Tongji University, China No. 1239 Siping Road, Shanghai 200092, China; lcf@tongji.edu.cn
6 Hebei Haoyu Engineering Technology Consulting Co., Ltd., No. 238 Jin Zhong He Street, Tianjin 300250, China; liuxiushui@126.com
7 School of Civil Engineering, Chongqing University, No. 83 Shabei Street, Chongqing 400045, China
* Correspondence: liufei1991c@163.com

Abstract: The behavior of buried offshore pipelines subjected to upheaval buckling has attracted much attention in recent years. Numerous researchers have made great efforts to investigate the influence of different soil cover depth ratios, soil strengths and pipe-soil interfaces on failure mechanisms and bearing capacities during pipeline uplift. However, attention to soil spatial variability has been relatively limited. To address this gap, a random small-strain finite element analysis has been conducted and reported in this paper to evaluate the influence of the random distribution of soil strength on pipe uplift response. The validity of the numerical model was verified by comparison with the results presented in the previous literature. The spatial variation of soil strength was simulated by a random field. The effect of soil variability on the failure mechanism was determined by comparing the displacement contours of each random realization. Probabilistic analyses were performed on the random uplift capacity obtained by a series of Monte Carlo simulations, and the relationship between the failure probability and the safety factor was also determined. The findings of the present work might serve as a reference for the safety designs of pipelines.

Keywords: pipelines; finite element analysis; random field; failure mechanism; uplift capacity

1. Introduction

Offshore oil and gas projects have experienced rapid growth in recent decades. Pipelines, as a safe, effective and swift transportation means, have been deployed widely for the conveyance of hydrocarbons from offshore fields to other processing facilities [1–3]. Generally, the pipelines are buried into the seabed to diminish external forces acting on them, which are caused by waves or human activities (see Figure 1). The transportation of fluids through the pipeline under very high temperatures and pressure conditions leads to thermal expansion, thus making it susceptible to buckling [4,5]. For buried pipes, upheaval buckling is a major current focus for oil and gas projects and, in this scenario, its resistance is affected by the shear strength and the submerged unit weight of the soil cover.

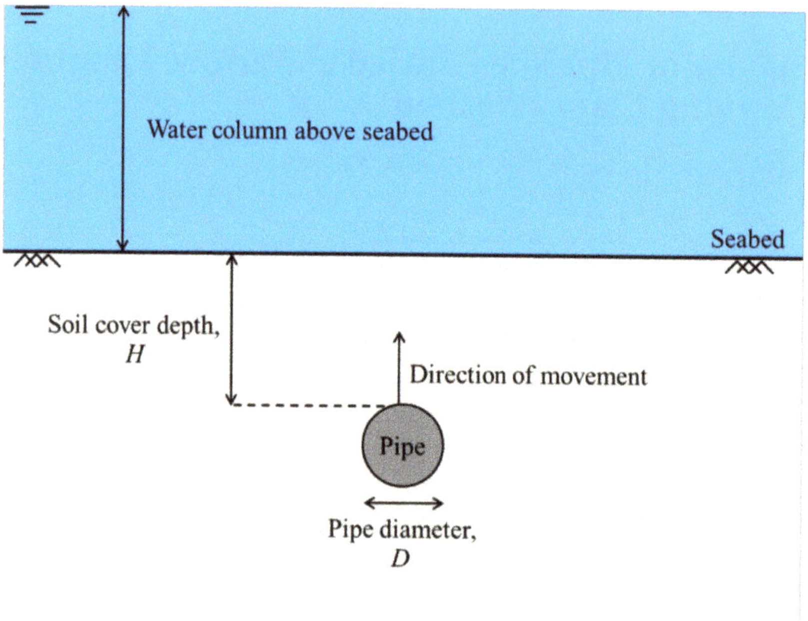

Figure 1. Problem geometry in this study.

Numerous works have been conducted to investigate the uplift capacity and failure mechanism for pipelines, including experimental works and numerical modeling [6–9]. In existing studies, the influence of soil weight, soil cover depth, pipeline-soil interface and other factors has been extensively investigated. For instance, Liu et al. [9] conducted vertical pullout tests on pipelines of different diameters and at different buried depths and proposed a nonlinear soil resistance model to simulate the bearing capacity of the pipeline. Brennan et al. [8] carried out a series of centrifuge experiments in bulk clay to evaluate the effect of backfill on the uplift capacity. Maitra et al. [7] performed small-strain finite element analyses to investigate the influence of soil cover depths, soil–pipeline interfaces and soil properties on the uplift behavior of pipelines in clay and proposed a generalized framework to evaluate pipeline uplift capacity. As for the failure mechanism of soil due to pipeline uplift, many scholars have pointed out that it can be divided into global soil failure and local soil failure [5]. A shallowly buried pipeline is prone to the former failure mode, in which the failure zone extends up to the seabed surface. As the soil cover depth increases, the latter failure mode occurs, characterized by soil flow around the pipeline. And there is a specific soil cover depth where failure mode transitions often take place.

Previous studies have undoubtedly contributed to readers' understanding of the upward buckling of pipelines. However, the soil in most of these analyses is treated as an idealized material, with homogeneous or linear increases in shear strength. In nature, the inherent properties of soil deposits generally vary spatially [10–18]. It has been widely recognized that such spatial variability in soil parameters can cause the mechanical behaviors of offshore structures to deviate from the results of deterministic analyses. Frequently, the spatial distributions of soil properties are treated as a random field and several random finite element analyses have been undertaken to explore the responses of offshore structures in spatially random soils. Charlton and Rouainia [19] undertook a probabilistic analysis on the bearing capacity of skirted foundations under combined vertical-horizontal-moment loadings and constructed several probabilistic failure envelopes. Cheng et al. [20] carried out a large deformation random element analysis to investigate the uplift responses of helical anchors in random clay, and the failure mechanisms and pullout capacities of anchors under different statistical characteristics are extensively analyzed. Luo and Li [21] combined large deformation finite element analysis, small-strain finite element analysis and random field theory to analyze the combined bearing capacities of spudcans considering installation effects in spatially varying soils.

In this paper, a random small-strain finite element (*SSFE*) analysis was conducted to study the uplift behavior of pipelines in a spatially varying clayey seabed. The validity of the numerical strategies adopted in the current study was verified by comparisons with previous numerical research. The soil's undrained shear strength was simulated as a random field and realized in the finite element software ABAQUS 6.14 in the form of field variables. The effect of the random distribution of soil strength on the failure mechanism and uplift capacity was determined by Monte Carlo simulations. Finally, for the random uplift capacity of pipelines buried at different depths, the relation between different failure probabilities and safety factors was determined. The results obtained in this paper may facilitate design guidelines to take into account spatial variability in soil properties.

2. Modeling of Spatially Variable Soil

As pointed out in past literature, the soil's undrained shear strength (i.e., s_u) significantly affects the failure mechanism of a pipeline as well as the uplift capacity, thus s_u was modeled as a random field in this study for representation of spatial variability. According to the studies of Yi et al. [22] and Cheng et al. [20], s_u can be considered to follow a normal or lognormal distribution. In this paper, the lognormal distribution was adopted to avoid the appearance of negative s_u values. Then, the statistical characteristics of the random variable can be quantified by the coefficient of variation (i.e., *COV*) and the correlation length (i.e., θ). With reference to Phoon and Kulhawy [10], the typical coefficient of variation of s_u (i.e., COV_{s_u}) ranges from 0.1 to 0.5. As such, the $COV_{s_u} = 0.3$ was chosen as a typical value. Furthermore, the correlation length in the horizontal direction (i.e., θ_h) is usually larger than the correlation length in the vertical direction (i.e., θ_v). Following Charlton and Rouainia [23], the correlation lengths of $\theta_h = 6D$ and $\theta_v = D$ were selected (where D denotes the pipeline diameter).

For purpose of carrying out random finite element analyses, several efficient methods have been developed for the generation of random fields. Examples of these methods are as follows: the spectral representation method, the Karhunen-Loeve expansion method, the linear estimation method and the modified linear estimation method (MLEM). The MLEM, proposed by Liu et al. [24], was employed in this study because of its efficiency.

3. Finite Element Methodology

3.1. Small-Strain Finite Element Analysis Model

Pipelines are often used for long-distance transport, resulting in a very large length–diameter ratio. Therefore, a small-strain finite element (*SSFE*) analysis, based on the plane strain assumption, was adopted in this study to simulate the pipeline uplift process in order to improve computational efficiency.

Figure 2 depicts the typical mesh and the boundary dimensions of the two-dimensional finite element model. To ensure minimal boundary influence on the uplift response of the pipeline, the distance between the outer side of the pipeline and the lateral sides of the model as well as the distance between the lower part of the pipeline and the bottom of the model were set to $5D$ [5]. Meanwhile, the lateral faces of the model were constrained in the perpendicular direction, and the bottom boundary was restricted in all directions.

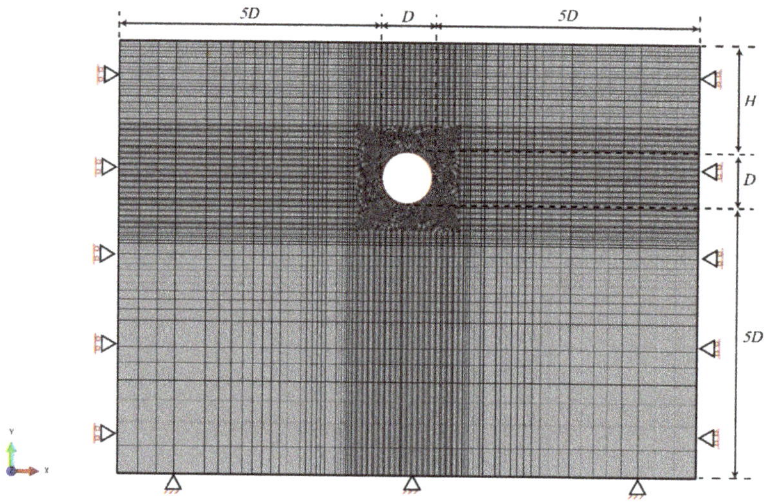

Figure 2. Exemplar finite element mesh and boundary conditions ($H/D = 2$).

The soil domain was discretized using 4-node bilinear plane strain quadrilateral elements (i.e., CPE4). A finer mesh with a size of 0.01%D was prescribed in the vicinity of the pipeline since such a mesh size was adopted by Maitra et al. [5]. The elastic-perfectly-plastic Tresca constitutive model was adopted to model the behavior of the soil. The clayey seabed was assumed to be undrained and the Poisson's ratio (ν) of 0.49 was employed [25]. Although an undrained condition implies a Poisson's ratio of 0.5, a smaller value, like the one used in this paper (0.49), promotes numerical stability and efficiency without affecting the computed uplift capacity of a pipeline. The Young's modulus of clay (E) was set to be 500 times the soil's undrained shear strength ($E = 500s_u$). Having considered the practical ranges documented in the existing literature, a pipeline of a diameter of 1 m was selected. The magnitudes of the parameters used to define the soil and the pipe are summarized in Table 1. The stiffness of a pipeline is markedly greater than that of clay, thus the pipeline was modeled as a rigid body in this study [26–28]. In addition, for numerical expediency, the pipeline–soil interface was considered as rough with an infinite interface tensile strength (i.e., rough full tension condition) and the Tie constraint in ABAQUS was used for modeling such contact, which avoided considering the pipeline–soil interface elements. Meanwhile, the wished-in-place assumption was adopted and the pipeline was embedded at various anticipated depths. The displacement-controlled approach was employed to model the uplift process. A reference point was specified on the top of the pipeline and a vertical displacement was applied to the reference point. Then, the uplift capacity of the pipeline was obtained by outputting the reaction force of the reference point.

Table 1. The magnitudes of the pipe and soil parameters.

Parameter	Symbol	Property Value	Units
Pipe diameter	D	1	m
Soil density	ρ	1600	kg/m^3
Poisson's ratio	ν	0.49	
Undrained shear strength	s_u	30	kPa
Young's modulus	E	$500s_u$	MPa

3.2. Validation of the Numerical Model

In order to verify the validity of the above-mentioned *SSFE* method, the uplift capacity factor N_u obtained in this paper was compared with the results obtained in earlier studies. Here, N_u can be calculated by the following equation:

$$N_u = V_u/(Ds_u) \quad (1)$$

where V_u is the ultimate uplift capacity.

Maitra et al. [5] carried out a series of small-strain finite element analyses to explore the uplift behavior of the pipeline in uniform clays. A range of soil cover ratios (i.e., H/D ratios, where H represents the initial soil cover depth of the pipeline.) between 0.5 and 6 were considered. Table 2 illustrates the comparisons of the N_u value for the rough pipelines under full tension. It can be seen that for pipelines at different depths, the calculated results from this study agree well with the work of Maitra et al. [5], with a maximum discrepancy of about 1%. Furthermore, when the soil cover depth ratio H/D was less than 2, N_u increased with the increase of H/D and reached a maximum value when H/D equaled 2 and remained constant as H/D increased, indicating a critical depth ratio for $H/D = 2$. According to the study of Kumar et al. [4], in the case of a full tension condition, $H/D = 2$ is the critical depth ratio for a failure mechanism to transition from a global failure to a local failure. The results of this study also indicate this trend. Furthermore, according to Randolph and Houlsby [29], for a lateral-loaded cylindrical pile with the local shear failure mechanism, the analytical solution of its bearing capacity factor is 11.94, while the ultimate bearing capacity factor obtained in this paper was 12.11. These results are in good agreement with each other. Therefore, it can be concluded that the *SSFE* method adopted in this paper can effectively simulate the uplift process of a pipeline.

Table 2. Comparison of uplift capacity factor between this study and previous numerical analysis.

H/D	N_u (Maitra et al. [5])	N_u (This Study)
0.5	9.02	8.98
1	10.47	10.59
2	11.98	12.11
3	11.98	12.11
4	11.98	12.11
5	11.98	12.11
6	11.98	12.11

3.3. Random SSFE Analysis

Starting from this section, the undrained shear strength of the clayey seabed was considered as a random field and the random *SSFE* analyses were conducted. Six H/D values (i.e., 0.5, 1, 2, 3, 4 and 5) were considered, which are representative of typical soil cover depth ratios. The mean value of the soil's undrained shear strength (i.e., μ_{s_u}) was set to be 30 kPa. It should be noted herein that the random *SSFE* analysis in this study was a univariate analysis. Considering that s_u was a random variable obeying lognormal distribution, there was a direct linear relationship between Young's modulus E and soil strength (i.e., $E = 500s_u$), resulting in E being a dependent random variable related to s_u. In addition, other model details for random *SSFE* analyses, such as boundary dimensions, mesh size, element type, etc., were the same as those noted in Section 3.1.

4. Results and Discussions

4.1. Failure Mechanism in Deterministic and Random Soil

The effect of the spatial variation of the soil's undrained shear strength on the failure mechanism of the soil surrounding the pipeline is presented in this section. Figure 3a–d plot the displacement contours of the soil around the pipeline with the soil cover ratio H/D of 0.5, 1, 2 and 3 in the deterministic soil with s_u = 30 kPa, respectively. It can be seen that in all the deterministic analyses, when pipeline upheaval buckling occurred, the displacement contours of the soil were symmetrical and the failure mechanism was closely related to the value of H/D. For pipelines with a H/D less than 2, the failure zone extended to the seabed surface, corresponding to the global failure mode. However, when H/D was greater than or equal to 2, a localized circumferential soil flow surrounding the pipeline occurred. This is consistent with the conclusions obtained by Kumar et al. [4], which suggested that critical soil cover depth ratios for both the global and local failure modes was 2 for rough pipelines under full tension in uniform soils.

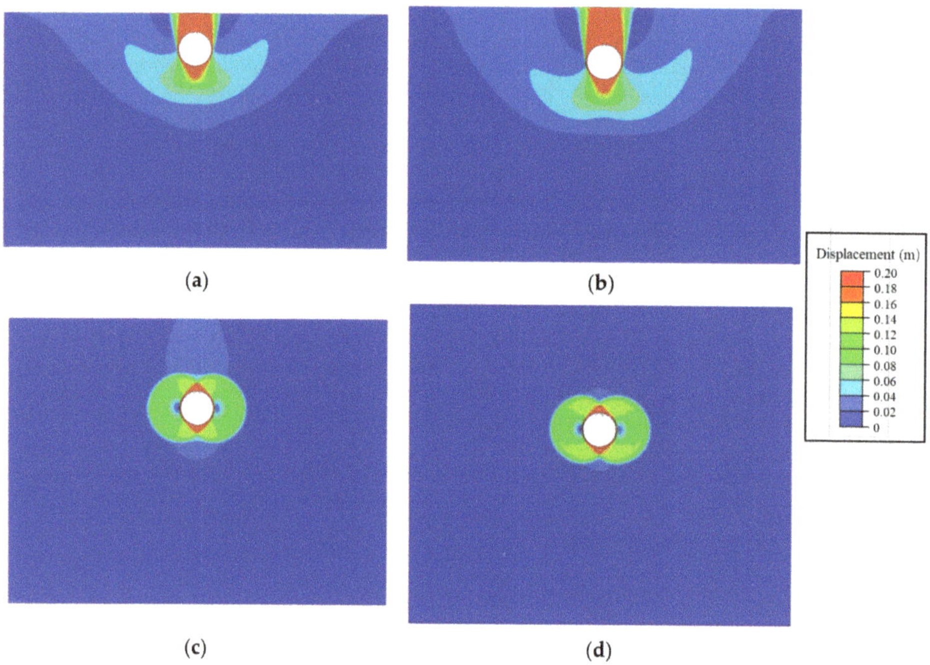

Figure 3. Uplift failure mechanisms for pipelines embedded at different depths in deterministic soils: (**a**) $H/D = 0.5$; (**b**) $H/D = 1$; (**c**) $H/D = 2$; (**d**) $H/D = 3$.

Figures 4 and 5 depict the displacement contours and the corresponding soil's undrained strength distribution in the random soil with pipeline soil cover depth ratios of 0.5 and 1, respectively. As can be seen from the figures, when the soil cover depth ratio was relatively small (i.e., $H/D = 0.5,\ 1$), the failure mechanism in the random realizations was similar to the results in the deterministic analysis. The displacement contour extended up to the seabed surface, presenting a global failure mode as a whole. Meanwhile, due to the random distribution of the soil's undrained shear strength, the displacement contours became asymmetric. Random bearing capacity factors (i.e., $N_{u,ran}$) deviated from the deterministic results (i.e., $N_{u,det}$), due to the existence of soil variability. In this study, $N_{u,ran}$ was calculated by the following equation:

$$N_{u,ran} = V_{u,ran}/(D\mu_{s_u}) \tag{2}$$

where $V_{u,ran}$ denotes the uplift capacity in random soils.

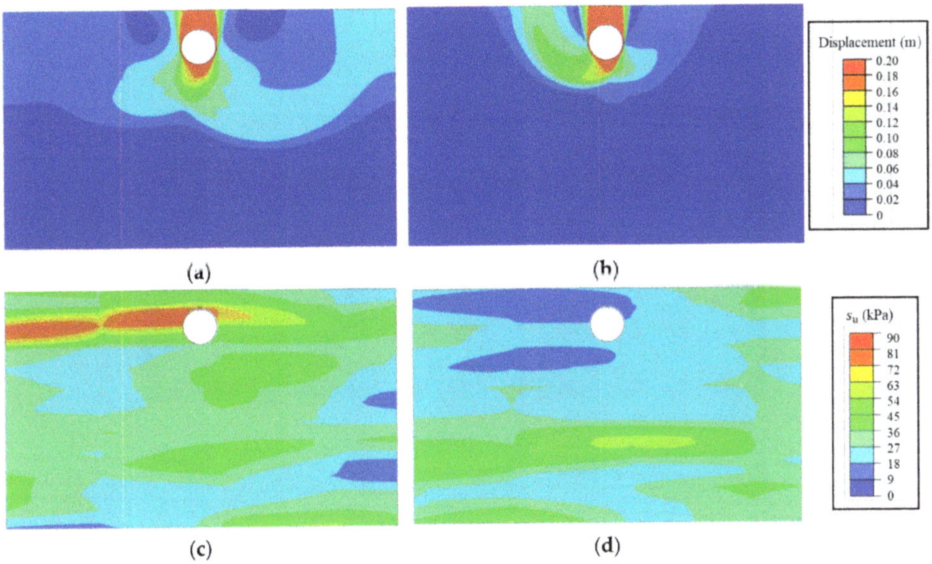

Figure 4. Displacement contours and corresponding soil's undrained strength of two random realizations at 0.5*D* depth: (**a**,**b**) displacement contours; (**c**,**d**) soil's undrained strength.

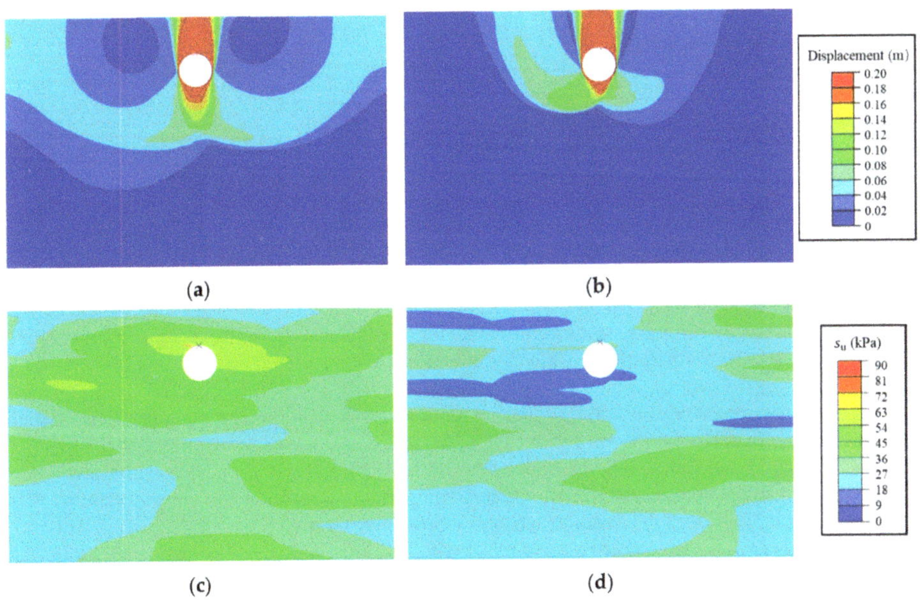

Figure 5. Displacement contours and corresponding soil's undrained strength of two random realizations at 1*D* depth: (**a**,**b**) displacement contours; (**c**,**d**) soil's undrained strength.

It should be noted that Figure 4a shows the displacement contour when the random uplift capacity factor of the pipeline with $H/D = 0.5$ reached its maximum and its soil strength distribution corresponded to that shown in Figure 4c. Furthermore, Figure 4b plots the displacement contour when the uplift capacity factor was the smallest and its soil strength distribution corresponded to that shown in Figure 4d. This correspondence also applies in the results shown in Figures 5–7. The influence of the spatial variability of s_u on the failure mechanism was illustrated by establishing the relationship between the uplift capacity factor, the failure zone and the strength of the soil around the pipeline.

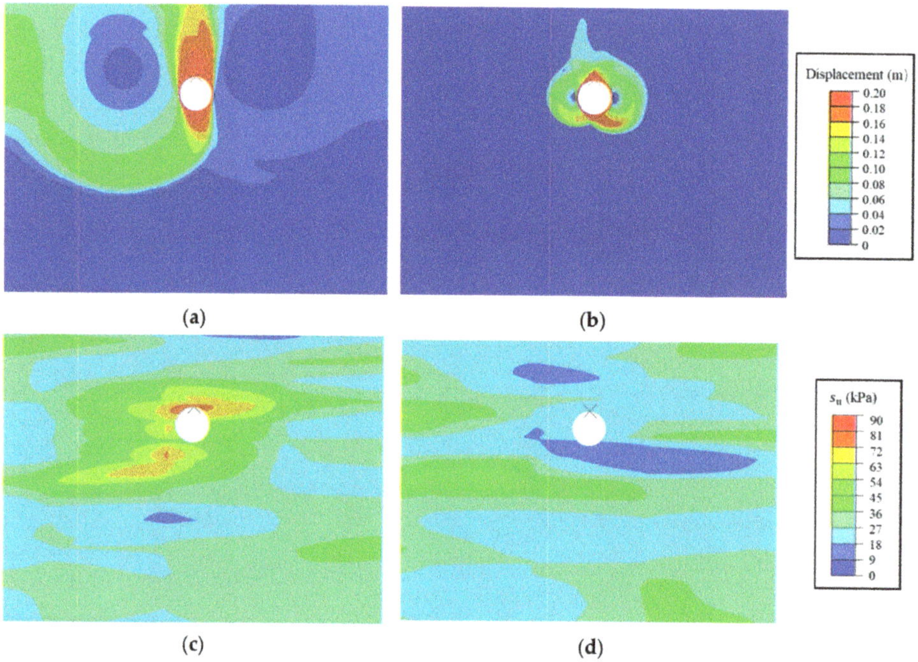

Figure 6. Displacement contours and corresponding soil's undrained strength of two random realizations at 2D depth: (**a**,**b**) displacement contours; (**c**,**d**) soil's undrained strength.

Taking the pipeline with $H/D = 0.5$ as an example, the failure zone in Figure 4a was larger than the deterministic result shown in Figure 3a and meanwhile, the soil strength around the pipeline was obviously greater than 30 kPa, which gave rise to the random uplift capacity factor ($N_{u,ran} = 13.48$) being significantly larger than the deterministic result ($N_{u,det} = 8.98$). By comparison, combined with Figure 4b,d, it is clear that a smaller failure zone and smaller soil strength around the pipe lead to a smaller random uplift capacity factor ($N_{u,ran} = 5.63$).

Figure 6 presents the displacement contours and corresponding soil strengths of pipelines with soil cover depth ratios of 2 in random soils. It was found that the failure mechanism in random soils was dependent on soil cover depth, but the results were different from the deterministic analysis due to the random distribution of soil strength. In uniform soil (ignoring soil variability), $H/D = 2$ was the critical soil cover depth ratio, that is, when H/D was greater than or equal to 2, the soil failure mode corresponded to the local failure mode. However, as shown in Figure 6a, when there existed a stronger interlayer around the pipeline, the displacement contour extended to the soil surface, showing a global failure mode. Conversely, as shown in Figure 6b, when the soil strength around the pipeline was weak, the failure mechanism corresponded to the local failure mode. This indicated that in random soils $H/D = 2$ was insufficient to ensure that the soil failure mode was the local failure one. For the pipeline with $H/D = 2$ in the spatially variable soil, the larger and smaller soil surrounding the pipeline made the failure modes to be global failure mode and local failure mode, respectively.

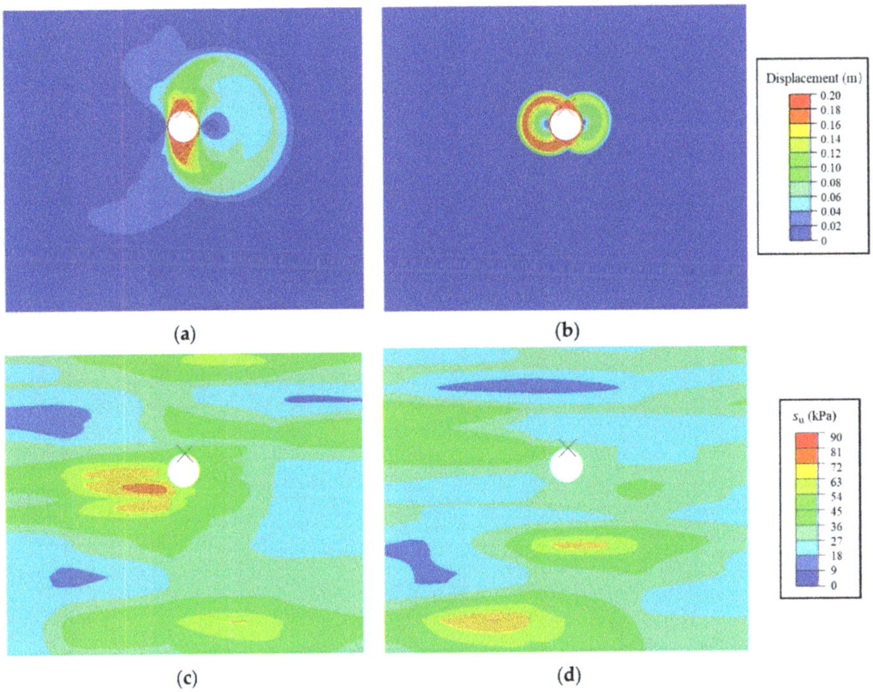

Figure 7. Displacement contours and corresponding soil's undrained strength of two random realizations at $3D$ depth: (**a**,**b**) displacement contours; (**c**,**d**) soil's undrained strength.

Figure 7 shows the displacement contours and corresponding soil strengths of pipelines with $H/D = 3$. It was found that no matter whether the strength of the soil surrounding the pipeline was strong or weak, the failure zone would not extend to the soil surface, that is, the failure mechanism was the local failure mode. This means that when the spatial variability of soil was considered, the critical soil cover depth ratio of the rough pipelines under full tension was 3. For the foundation of different buried depths in spatially random soils, Li et al. [30] pointed out that the critical value of the buried depth ratio was 3, which agrees well with the results presented in this paper.

4.2. Uplift Capacity Factors for Pipelines at Various Depths

In the present study, Monte Carlo simulations were carried out for each random realization of soil strength for pipelines with different soil cover depth ratios. Then the statistical characteristics of random uplift capacity factors were determined. Generally, the errors of the Monte Carlo approach decrease with an increase in the number of simulations n_{sim}. Thus, simulation accuracy can only be guaranteed once n_{sim} is large enough. Figure 8 displays the results of the convergence analyses for the number of simulations required for pipelines with different soil cover depth ratios. It was found that in all random *SSFE* analyses, the mean and standard deviation of $N_{\mathrm{u,ran}}$ converged over 200 simulations. Given that six soil cover depth ratios were considered in this paper, a total of 1200 random realizations of soil's undrained shear strength were performed for Monte Carlo simulations.

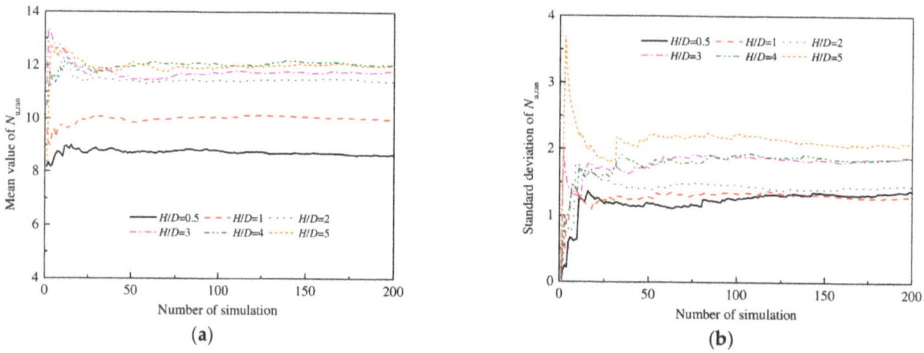

Figure 8. Variation trend for the mean and standard deviation of $N_{u,ran}$ with the number of Monte Carlo simulations. (**a**) Mean of $N_{u,ran}$. (**b**) Standard deviation of $N_{u,ran}$.

Figure 9 shows the uplift capacity factors of pipelines buried at different depths in deterministic and random soils. Also plotted in the figure is the mean value of the random uplift capacity factor (i.e., $\mu_{N_{u,ran}}$). It was found that, due to consideration of the spatial variability of the soil strength, the bearing capacity factor was distributed in a large range, fluctuating between 5.5 and 18. Moreover, similar to the deterministic analyses, $\mu_{N_{u,ran}}$ increased as the soil cover depth ratio (H/D) increased and tended to be a constant value after H/D reached the critical depth ratio (i.e., $H/D = 3$), which was due to the transition from a shallow global failure mode to a deeply buried local failure mode. At the same time, for pipelines in random soils, $\mu_{N_{u,ran}}$ was smaller than that of the results of the deterministic analysis at the same soil cover depth ratio. This phenomenon has also been observed in other random finite element analyses, which can be attributed to the fact that once soil variability is taken into account, the failure zone tends to follow the weakest path, thereby reducing the bearing capacity [30].

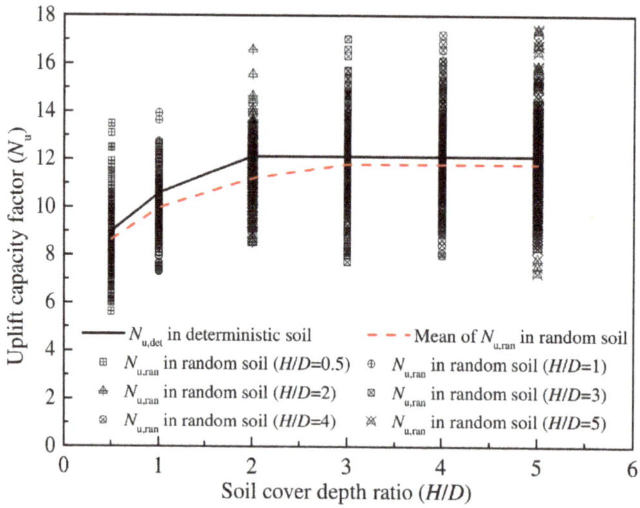

Figure 9. Uplift capacity factors of pipelines buried at various depths in deterministic and random soils.

Based on the above discussion, it can be suggested that the random distribution of the undrained shear strength of the soil significantly affected the uplift capacity factor of the pipeline, so it was necessary to carry out probabilistic analyses on its bearing capacity. For the random uplift capacity factor $N_{u,ran}$ of different soil cover depth ratios, the goodness-of-fit test was carried out. In this paper, the Kolmogorov–Smirnov (K–S) normal distribution goodness-of-fit test was used to verify whether $N_{u,ran}$ conformed to the lognormal distribution. Table 3 lists the mean and coefficient of variation of $N_{u,ran}$ (i.e., $\mu_{N_{u,ran}}$ and $COV_{N_{u,ran}}$) at various depths and the calculated p-values. As can be seen, all p-values were greater than 0.05, indicating that $N_{u,ran}$ followed the lognormal distribution [31]. Moreover, Figure 10 shows the histogram of the uplift capacity factor at different depths and the corresponding lognormal distribution fitting curve.

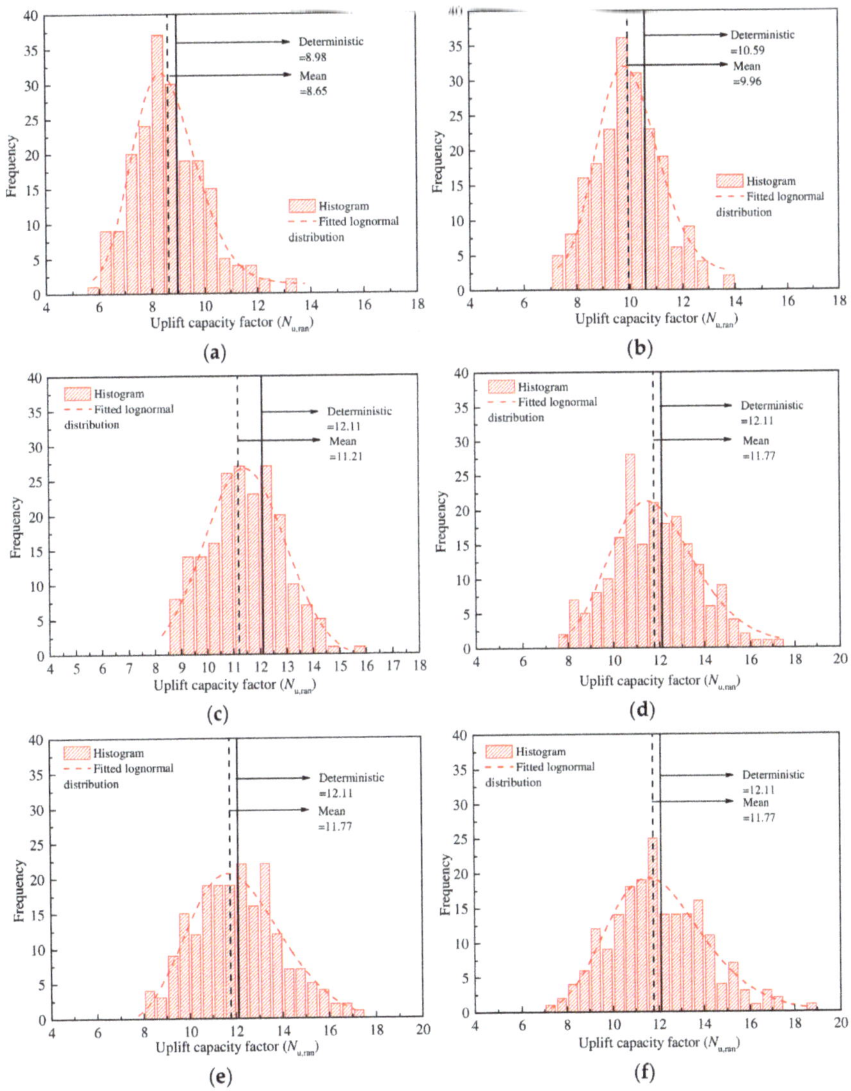

Figure 10. Histogram and fitted lognormal distribution of pipelines with embedment ratio H/D of: (**a**) 0.5; (**b**) 1; (**c**) 2; (**d**) 3; (**e**) 4; (**f**) 5.

Table 3. Uplift capacity factor statistics in random soil and the K–S test results.

H/*D*	$\mu_{N_{u,ran}}$	$COV_{N_{u,ran}}$	*p*-Values
0.5	8.65	0.13	0.88
1	9.96	0.15	0.61
2	11.21	0.16	0.72
3	11.77	0.16	0.92
4	11.77	0.16	0.86
5	11.77	0.17	0.83

Regarding the bearing capacity of foundations in spatially variable soils, Griffiths and Fenton [32] suggested that failure can be defined as a situation where the random bearing capacity is less than the deterministic bearing capacity. This definition was also adopted in the current study—the failure was able to be expressed as when α was less than 1, where $\alpha = N_{u,ran}/N_{u,det}$. Then, the failure probability was calculated by the cumulative probability of α. Figure 11 shows the cumulative probability of α under different soil cover depth ratios. It was seen that the probability of less than 1 was more than 50%, which means that ignoring the spatial variability of soil led to a high failure probability. For this reason, a safety factor (*SF*) was introduced to reduce the deterministic bearing capacity and the failure probability was defined as the probability that the random uplift capacity factor ($N_{u,ran}$) was less than the reduced capacity factor ($N_{u,det}/SF$). Since $N_{u,ran}$ followed a lognormal distribution, the failure probability can be calculated as follows:

$$p(N_{u,ran} < N_{u,det}/SF) = \Phi\left[\frac{\ln(N_{u,det}/SF) - \mu_{\ln N_{u,ran}}}{\sigma_{\ln N_{u,ran}}}\right] \tag{3}$$

where Φ denotes the cumulative function of normal distribution, $\mu_{\ln N_{u,ran}}$ and $\sigma_{\ln N_{u,ran}}$ respectively denote the mean and standard deviation of the natural logarithm of $N_{u,ran}$.

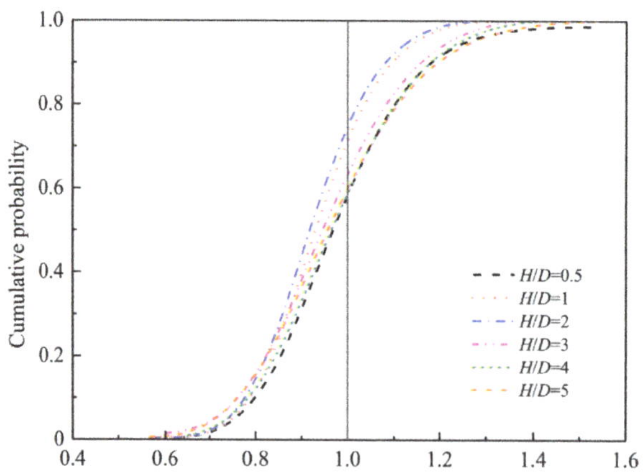

Figure 11. Cumulative probability of α for pipelines with different embedment ratios.

Table 4 summarizes the corresponding failure probability expressed in percentage for different safety factors. It was observed that the failure probability decreased sharply with the increase of the safety factor. In engineering applications, the failure probability is usually set at 1–0.1%, to ensure the safety of the structure [30]. For this reason, the safety factor corresponding to the failure probability below 1% was calculated and the results are shown in Table 5. It was seen that for pipelines with different soil cover depth ratios, a safety factor of 1.6 would ensure that the failure probability was less than 1%.

Table 4. Failure probabilities for different SF and soil cover depth ratios.

SF	*H/D*					
	0.5	1	2	3	4	5
1	52.95	70.45	74.49	63.87	58.97	59.29
1.2	13.76	19.06	21.83	22.02	17.14	20.76
1.4	1.89	1.92	2.31	4.24	2.60	4.43
1.6	0.17	0.09	0.12	0.54	0.25	0.67
1.8	0.01	/	/	0.05	0.02	0.08
2	/	/	/	/	/	0.01

Note: / represents that failure probability is less than 0.01%.

Table 5. Safety factors for different failure probabilities and soil cover depth ratios.

Failure Probability (%)	*H/D*					
	0.5	1	2	3	4	5
1	1.46	1.45	1.46	1.54	1.49	1.56
0.1	1.64	1.60	1.61	1.75	1.67	1.78
0.01	1.77	1.70	1.71	1.89	1.81	1.94

5. Conclusions

A two-dimensional random small-strain finite element analysis was carried out to study the uplift behavior of pipelines buried at various depths considering soil strength spatial variability. First, the reliability of the finite element model used in this paper was verified by comparisons with the results of past research. Then, the soil strength was simulated as a random field and a random finite element analysis was conducted using the Monte Carlo framework. Finally, the failure mechanism of each random realization of soil strength was compared and the probability analysis of the random uplift bearing capacity was performed. The major observations can be listed as follows:

(1) For rough pipelines under full tension, the random distribution of soil strength significantly affected its upward upheaval failure mechanism. When the spatial variability of soil was considered, the critical soil cover depth ratio corresponding to the transition of the failure mode from global to local failure was 3.

(2) For pipelines with different soil cover depth ratios, ignoring the soil variability would lead to a failure probability of more than 50%, resulting in an unconservative design.

(3) The failure probability would decrease sharply with the increase of the safety factor and a safety factor of 1.6 was required to satisfy the failure probability of less than 1%. This means that when considering the spatial variability of soil strength, a design guideline can specify a safety factor of 1.6 to reduce the deterministic results for achieving a target failure probability.

It is important to note that this study mainly focused on the uplift behavior of pipelines under different soil cover depth ratios when considering the spatial variability of soil strength and the conclusions obtained are only applicable to rough pipelines under full tension in specific soil strength configurations. The response to the upheaval buckling of pipelines was also affected by other factors, such as soil weight and the pipe–soil interface, etc., which require further research.

Author Contributions: Conceptualization, P.C. and F.L.; Formal analysis, C.L. and X.L.; Funding acquisition, K.Y.; Investigation, J.G.; Methodology, P.C., J.G., C.L. and F.L.; Resources, X.L. and K.Y.; Software, P.C.; Supervision, K.Y.; Validation, P.C.; Writing—original draft, P.C. and F.L. All authors have read and agreed to the published version of the manuscript.

Funding: This research was funded by the Shandong Excellent Young Scientists Fund Program (Overseas) (2022HWYQ-016), Shandong Provincial Natural Science Foundation (ZR2021QE254) and Guangdong Basic and Applied Basic Research Foundation (2021A1515110564).

Institutional Review Board Statement: Not applicable.

Informed Consent Statement: Not applicable.

Data Availability Statement: Not applicable.

Conflicts of Interest: The authors declare no conflict of interest.

References

1. Sheil, B.B.; Byrne, B.W.; Martin, C.M. Rate effects on the uplift capacity of pipelines embedded in clay: Finite element modelling. *Comput. Geotech.* **2021**, *137*, 104155. [CrossRef]
2. Maitra, S.; Chatterjee, S.; White, D.; Choudhury, D. Uplift resistance of buried pipelines: The contribution of seepage forces. *Ocean. Eng.* **2022**, *250*, 111037. [CrossRef]
3. Seth, D.; Manna, B.; Kumar, P.; Shahu, J.T.; Fazeres-Ferradosa, T.; Taveira-Pinto, F.; Carvalho, H. 2021, Uplift and lateral buckling failure mechanisms of offshore pipes buried in normally consolidated clay. *Eng. Fail. Anal.* **2022**, *121*, 105161. [CrossRef]
4. Kumar, P.; Seth, D.; Manna, B.; Shahu, J.T. Lateral and Uplift Capacity of Pipeline Buried in Seabed of Homogeneous Clay. *J. Pipeline Syst. Eng. Pract.* **2021**, *12*, 04021020. [CrossRef]
5. Maitra, S.; Chatterjee, S.; Choudhury, D. Generalized framework to predict undrained uplift capacity of buried offshore pipelines. *Can. Geotech. J.* **2016**, *53*, 1841–1852. [CrossRef]
6. Chen, R.P.; Zhu, B.; Ni, W.J. Uplift tests on full-scale pipe segment in lumpy soft clay backfill. *Can. Geotech. J.* **2016**, *53*, 578–588. [CrossRef]
7. Martin, C.M.; White, D.J. Limit analysis of the undrained bearing capacity of offshore pipelines. *Géotechnique* **2012**, *62*, 847–863. [CrossRef]
8. Brennan, A.J.; Ghahremani, M.; Brown, M.J. Strength reduction for upheaval buckling of buried pipes in blocky clay backfill. *Ocean. Eng.* **2017**, *130*, 210–217. [CrossRef]
9. Liu, R.; Basu, P.; Xiong, H. Laboratory tests and thermal buckling analysis for pipes buried in Bohai soft clay. *Mar. Struct.* **2015**, *43*, 44–60. [CrossRef]
10. Phoon, K.K.; Kulhawy, F.H. Characterization of geotechnical variability. *Can. Geotech. J.* **1999**, *36*, 612–624. [CrossRef]
11. Phoon, K.K.; Kulhawy, F.H. Evaluation of geotechnical property variability. *Can. Geotech. J.* **1999**, *36*, 625–639. [CrossRef]
12. Liu, Y.; Zhang, L. Seismic response of pile–raft system embedded in spatially random clay. *Géotechnique* **2019**, *69*, 638–645. [CrossRef]
13. Liu, Y.; Zhang, W.; Zhang, L.; Zhu, Z.; Hu, J.; Wei, H. Probabilistic stability analyses of undrained slopes by 3D random fields and finite element methods. *Geosci. Front.* **2018**, *9*, 1657–1664. [CrossRef]
14. Liu, Y.; He, L.Q.; Jiang, Y.J.; Sun, M.M.; Chen, E.J.; Lee, F.H. Effect of in situ water content variation on the spatial variation of strength of deep cement-mixed clay. *Géotechnique* **2019**, *69*, 391–405. [CrossRef]
15. Liu, Y.; Jiang, Y.J.; Xiao, H.; Lee, F.H. Determination of representative strength of deep cement-mixed clay from core strength data. *Géotechnique* **2017**, *67*, 350–364. [CrossRef]
16. Liu, Y.; Lee, F.H.; Quek, S.T.; Chen, E.J.; Yi, J.T. Effect of spatial variation of strength and modulus on the lateral compression response of cement-admixed clay slab. *Géotechnique* **2015**, *65*, 851–865. [CrossRef]
17. Chen, X.; Li, D.; Tang, X.; Liu, Y. A three-dimensional large-deformation random finite-element study of landslide runout considering spatially varying soil. *Landslides* **2021**, *18*, 3149–3162. [CrossRef]
18. Chen, X.J.; Fu, Y.; Liu, Y. Random finite element analysis on uplift bearing capacity and failure mechanisms of square plate anchors in spatially variable clay. *Eng. Geol.* **2022**, *304*, 106677. [CrossRef]
19. Charlton, T.S.; Rouainia, M. A probabilistic approach to the ultimate capacity of skirted foundations in spatially variable clay. *Struct. Saf.* **2017**, *65*, 126–136. [CrossRef]
20. Cheng, P.; Liu, Y.; Li, Y.P.; Yi, J.T. A large deformation finite element analysis of uplift behaviour for helical anchor in spatially variable clay. *Comput. Geotech.* **2022**, *141*, 104542. [CrossRef]
21. Luo, W.; Li, J. Effects of installation on combined bearing capacity of a spudcan foundation in spatially variable clay. *Appl. Ocean. Res.* **2022**, *120*, 103055. [CrossRef]
22. Yi, J.T.; Huang, L.Y.; Li, D.Q.; Liu, Y. A large-deformation random finite-element study: Failure mechanism and bearing capacity of spudcan in a spatially varying clayey seabed. *Géotechnique* **2020**, *70*, 392–405. [CrossRef]
23. Charlton, T.S.; Rouainia, M. Probabilistic analysis of the uplift resistance of buried pipelines in clay. *Ocean. Eng.* **2019**, *186*, 105891. [CrossRef]
24. Liu, Y.; Lee, F.H.; Quek, S.T.; Beer, M. Modified linear estimation method for generating multi-dimensional multi-variate Gaussian field in modelling material properties. *Probabilistic Eng. Mech.* **2014**, *38*, 42–53. [CrossRef]
25. Taiebat, H.A.; Carter, J.P. Numerical studies of the bearing capacity of shallow foundations on cohesive soil subjected to combined loading. *Géotechnique* **2000**, *50*, 409–418. [CrossRef]
26. Liu, F.; Yi, J.; Cheng, P.; Yao, K. Numerical simulation of set-up around shaft of XCC pile in clay. *Geomech. Eng.* **2020**, *21*, 489–501.
27. Peng, Y.; Liu, H.; Li, C.; Ding, X.; Deng, X.; Wang, C. The detailed particle breakage around the pile in coral sand. *Acta Geotech.* **2021**, *16*, 1971–1981. [CrossRef]
28. Yi, J.T.; Liu, F.; Zhang, T.B.; Yao, K.; Zhen, G. A large deformation finite element investigation of pile group installations with consideration of intervening consolidation. *Appl. Ocean. Res.* **2021**, *112*, 102698. [CrossRef]

29. Randolph, M.F.; Houlsby, G.T. The limiting pressure on a circular pile loaded laterally in cohesive soil. *Géotechnique* **1984**, *34*, 613–623. [CrossRef]
30. Li, J.; Tian, Y.; Cassidy, M.J. Failure mechanism and bearing capacity of footings buried at various depths in spatially random soil. *J. Geotech. Geoenviron. Eng.* **2015**, *141*, 04014099. [CrossRef]
31. Cheng, P.; Yi, J.T.; Liu, F.; Dong, J.J. Characterization of Model Uncertainty for the Vertical Pullout Capacity of Helical Anchors in Cohesive Soils. *J. Mar. Sci. Eng.* **2020**, *8*, 738. [CrossRef]
32. Griffiths, D.V.; Fenton, G.A. Bearing capacity of spatially random soil: The undrained clay Prandtl problem revisited. *Géotechnique* **2001**, *51*, 351–359. [CrossRef]

sustainability

MDPI

Article

The Influence of Mixing Degree between Coarse and Fine Particles on the Strength of Offshore and Coast Foundations

Shunkai Liu [1,2], Yuxing Nie [3], Wei Hu [1,4], Mohammed Ashiru [2,5], Zhong Li [6,*] and Jun Zuo [7]

1 School of Civil Engineering, Hunan University of Science and Technology, Xiangtan 411201, China; lsk1019@csu.edu.cn (S.L.); yilukuangben@hnust.edu.cn (W.H.)
2 School of Civil Engineering, Central South University, Changsha 410075, China; mohammeda@abu.edu.ng
3 School of Civil Engineering, Lanzhou Jiaotong University, Lanzhou 730070, China; 180308314@stu.lzjtu.edu.cn
4 Hunan Province Key Laboratory of Geotechnical Engineering Stability Control and Health Monitoring, Hunan University of Science and Technology, Xiangtan 411201, China
5 Department of Civil Engineering, Ahmadu Bello University, Zaria 810107, Nigeria
6 Department of P.E., Central South University, Changsha 410075, China; 712058@csu.edu.cn
7 China State Construction Railway Investment & Engineering Group Co., Ltd., Beijing 102600, China; liu865278280@126.com
* Correspondence: 712058@csu.edu.cn

Abstract: The variability in strata of foundation soil in marine environments makes it tedious to design foundations for offshore structures. Hence, it is essential to investigate and evaluate the strength properties of this type of soil. This study investigates the variability of the soil strata (which is quantified by the index of the mixing degree between coarse and fine particles) and its influence on the stability of the soil by mixing coarse and fine particles at varying proportions. A series of discrete element method triaxial shear tests were conducted on binary geotechnical mixtures with a varying proportion of coarse content (25%, 50% and 75%) and different mixing degrees (ranging from 0.0 to 1.0). The macroscopic results show that the peak shear strength increases with an increase in mixing degree, and the increase is more obvious with increasing coarse content, while the critical shear strength is independent of the mixing degree. The main evaluation of the number, mean normal force and distribution of the coarse–fine (cf) contact helps to clarify the meso-mechanisms that result in the variations in peak shear strength and critical shear strength with mixing degree. The increase in the peak strength may primarily be due to the increased number and globalized distribution of coarse–fine contact. However, the decreased contact force of coarse–fine contact counterbalances the strength gain due to the increased number and globalized distribution, which maintain the stability of the critical strength.

Keywords: granular mixtures; meso-mechanism; mixing degree; shear strength

1. Introduction

Binary geotechnical mixtures are ubiquitous in nature, especially in marine geotechnical engineering. As the direct support of offshore or coastal facilities, it is important to consider the stability of binary geotechnical mixtures during the construction and operation of offshore and coastal installations [1]. For example, as one of the main foundation forms of marine structures, the bearing capacity of the soil around the pile must be considered in the calculation and design of the screw anchor [2]. However, the formation and evolution of binary geotechnical mixtures offshore or in coastal regions is a natural morphodynamic process, predominantly affected by currents, tides, waves, and even storm surges [3,4]. The complex formation process results in distinct stratification of the offshore and coastal soils, i.e., uneven mixing between coarse and fine particles. For example, the stratigraphical results from the Shetland Islands show distinct stratification of coarse and fine particles, which is confirmed to have been caused by three tsunamis between 8000 and 1500 years ago [5]; some artificial activities also create layered coastal foundations [6]. Previous studies have shown that the uniformity of mixing between particles affects the local and even the global properties of granular materials [7,8]. However, this effect has not been systematically studied in the aspect of foundation stability in marine geotechnical engineering. In other fields of geotechnical engineering, some researchers have begun to pay attention to the influence of the interparticle mixing degree on the strength of the mixture. Azéma et al. [7] analyzed the effect of the mixing degree on shear strength using the discrete element method (DEM). They reported that the shear strength was not affected by the mixing degree. Although, it was observed that they considered the mixing of only disks and rods of the same diameter in 2D, which cannot represent the actual difference in size between rock and soil particles observed in the binary geotechnical mixtures. Dong et al. [9] performed a strength analysis for a soil–rock mixture by preparing three specimens at different mixing proportions. However, they did not give the quantitative description of the mixing proportions; hence, their conclusions could not be applied to other projects.

Therefore, the variability of the soil strata was quantified by the index of mixing degree between coarse and fine particles. And the mixing degree index proposed by Liu et al. [10], which can be obtained by the simple analysis of borehole sampling results, was used in this study to guide the preparation of specimens in this paper. The strength analysis of gap-graded binary geotechnical mixtures under different mixing degrees was carried out using a verified DEM model, which is an effective way to study the mechanical and structural behaviors of binary geotechnical mixtures from a micromechanical perspective [11,12]. The macroscopic trend and meso-mechanism of the variation in strength with mixing degree was analyzed intuitively.

2. Index of Mixing Degree and DEM Model Verification Result

The calculation principle of the mixing degree index and the verification process of the DEM model have been described in detail by the authors in [10], which is more to validate the usefulness of mixing degree and the DEM model. The research object in [10] is for a loose sample generated by the limitation of experimental instruments, and is completely different from the compacted sample used to analyze the stability of subgrade in this paper. Therefore, only these two results are concisely described here.

2.1. Index of Mixing Degree

The greatest shortcoming of the existing indices is that obtaining particle information is difficult to realize in marine geotechnical engineering; for example, all of the particle positions [13], the local particle concentration [14], particle number ratio [15], relative position ratio [16], contact number ratio [17], and bed height ratio [18]. In marine geotechnical engineering, the most common way to obtain particle information is through borehole exploration, which can indicate the distribution of different particles along the depth direction. Thus, the most applicable mixing index for this field should be based on the information obtained from borehole exploration.

Hence, after the improvement based on the index of Dai et al. [19], a new index is proposed and illustrated in combination with Figure 1. Figure 1 shows the content of particle A (A represents one kind of particle in a binary mixture) as a function of normalized depth H/HM (H is the actual height when the content of particle A is Vf, and HM is the binary mixture height). When the content of particle A is assumed to be Vf_0, Line 1 and Line 2 represent the two states of complete separation (Md = 0) and homogeneous mixing (Md = 1), respectively. Line X arbitrarily represents a mixing degree between 0 and 1. As long as a point on Line X does not perfectly coincide with Line 2, it is in the state of segregation, and the distance from this point to Line 2 directly reflects the degree of segregation at its corresponding height. Based on this principle, the improved mixing degree index adopted in this paper can be represented as follows:

$$Md = 1 - \frac{A_s}{A_i + A_{ii}} \quad (1)$$

where A_s is the area enclosed by Line 2 and Line X, and A_i and A_{ii} represent the area of the upper left and lower right rectangles enclosed by Line 1, Line 2 and the coordinate axes, respectively.

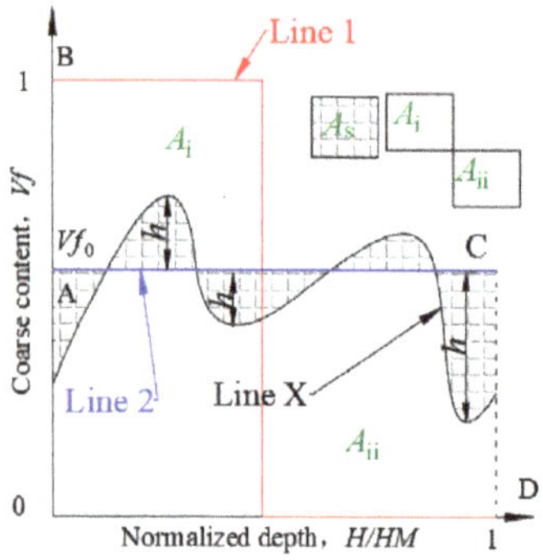

Figure 1. Calculation diagram of mixing degree.

Several researchers have focused on the relationship between coarse particle content and shear strength of binary geotechnical mixtures [20–22]. The results show that when the coarse particles content of a binary geotechnical mixture is outside the range of 25–75%, the physical and mechanical properties of the binary geotechnical mixture are usually affected by only the fine or coarse particles. Therefore, the coarse content of the tested binary geotechnical mixture specimens were 25%, 50% and 75%. In addition, combined with the following test dimensions, the specimens tested using the numerical simulations in this paper were divided into six layers, on average, for preparation. The coarse content as a function of normalized depth *H/HM* and the prepared specimens are as shown in Figure 2.

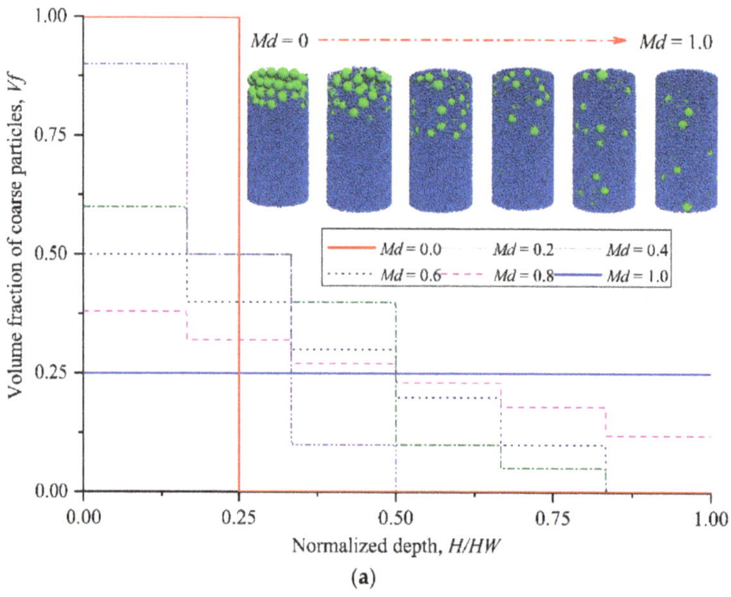

Figure 2. *Cont.*

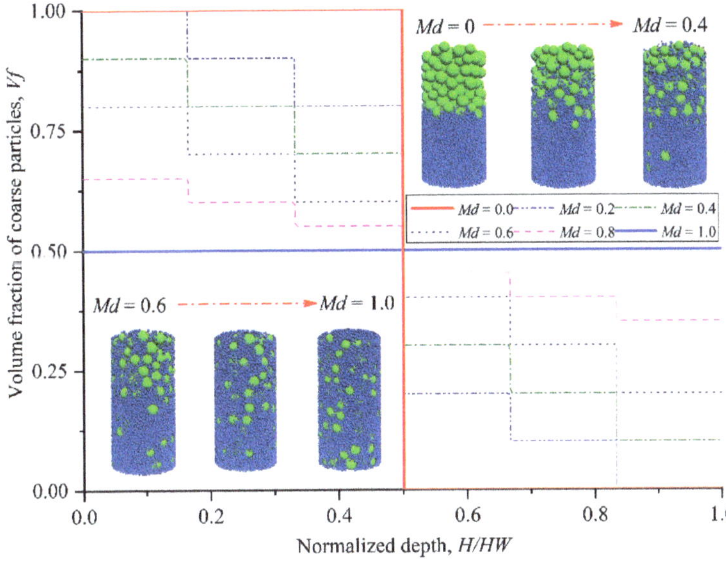

(**b**)

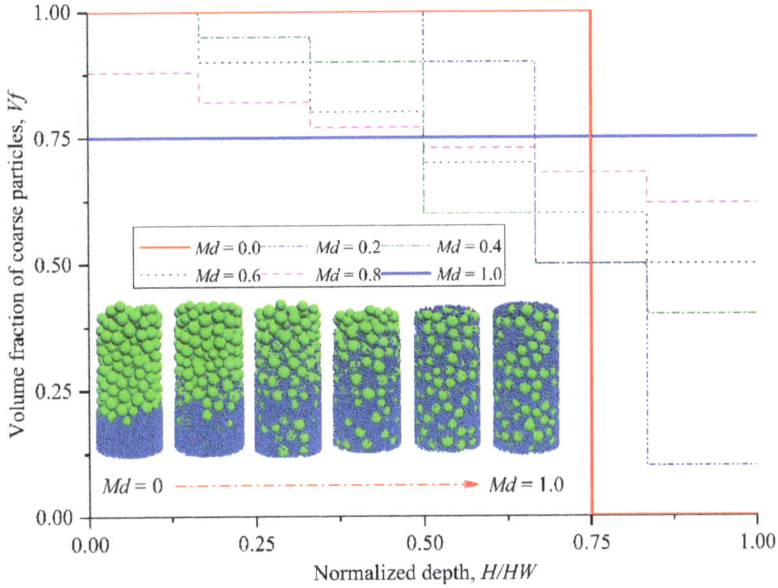

(**c**)

Figure 2. The coarse content as a function of normalized depth *H/HM* for the prepared specimens: (**a**) Vf_0 = 25%; (**b**) Vf_0 = 50%; and (**c**) Vf_0 = 75%.

2.2. DEM Model Verification Result

The reliability of the numerical simulation results were verified using PFC^{3D} to reproduce large-scale triaxial tests. The samples were prepared from particle packings with different coarse particle contents (25%, 50% and 75%) and *Md* values (0.0, 0.2, 0.4, 0.6, 0.8 and 1.0). Both the simulation and physical experiments were carried out with the same specimen size, which was $\Phi \times H$ = 300 mm × 600 mm (where Φ is the diameter and H is the height). The particles in the numerical simulation (NS) were spheres, and the particles used in the physical experiments (PE) were Yuhua pebbles, which have relatively smooth surfaces. According to previous numerical simulation findings, the particle size enlargement method has little influence on the mechanical behavior of the particle materials [23,24]. Therefore, considering both the size effect and calculation efficiency, the particle sizes assumed in the numerical simulations were doubled compared with those of the physical experiments. In the process of preparing samples for the physical experiments and numerical simulations, first, a mixture was made according to the *Md* index in Figure 2, which determines the coarse and fine particle contents in each layer. Then 18 samples are prepared from particle packings with three coarse particle contents (25%, 50% and 75%) and six *Md* values (0.0, 0.2, 0.4, 0.6, 0.8 and 1.0), respectively. The sample preparation process and quantity in Section 3 are similar. In the numerical simulation, the friction coefficient between particles was set as 0.6, which is consistent with the real particles and the calibration results in the physical experiment. This was also the main reason why the samples were relatively loose. The comparison results are as shown in Figure 3. For convenience, each sample is identified by test category and coarse content. For example, 75%-PE and 25%-NS represent physical experiment samples with a coarse content of 75% and numerical simulation samples with a coarse content of 25%, respectively.

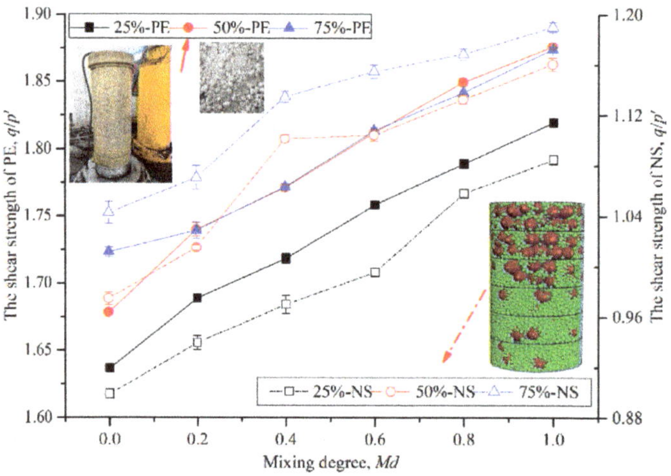

Figure 3. Reliability verification of the numerical simulation.

Figure 3 shows that the shear strength q/p' (where q is the deviatoric stress and p' is the effective mean stress) increases with an increase in the mixing degree index. It can be observed that the increasing trends for both the physical experiment and numerical simulation results at the same mixing degree index are highly consistent. Although the q/p' values in the physical experiments are higher than those in the numerical simulations by approximately 0.6–0.8, the result is acceptable when the shape factor is taken into account [22–26]. This is because the physical experiments used real particles, whose shapes deviate from spherical particles, while the numerical simulations use ideal spherical particles. Relevant studies have also proven that when the shape index of certain aggregate particles deviate from the spherical shape, generally, its shear strength is approximately 50% higher than that of spherical particles, with a maximum of 80% [25–29]. Therefore, it was appropriate to use these parameters in this paper to carry out numerical simulation and analyze the meso-mechanism of the influence of mixing degree on the shear strength of the studied binary geotechnical mixtures.

3. Specimen's Properties and Simulation Parameters

3.1. Specimen's Properties

The size and other parameters of the simulated specimens were exactly the same as those in the verification test, and the details are as follows. According to the research results of Medley [20] and Xu et al. [30], the soil/rock threshold was divided according to the limit of $0.05L_c$ (L_c is the minimum specimen size, normally 300 mm). Therefore, particles with a diameter of more than 15 mm were defined as "rock", which are referred to as "coarse particles" in this paper, and particles with a diameter of less than or equal to 15 mm were defined as "soil", which are referred to as "fine particles" in this paper. Therefore, considering both the size effect and calculation efficiency, the diameter of the coarse particles was 40–60 mm, and the diameter of the fine particles was 5–15 mm. Because the fine and coarse particle sizes were evenly distributed in their respective ranges, the average particle size ratio of the coarse and fine particles was 5, meeting the size requirements of crystallography. That is, when a fine particle could exactly fit in the void formed by four tangent coarse particles of the same size, the size ratio between them is 4.45 [31].

The preparation of a specimen was divided into three steps: (1) Six small closed cylinders with sizes $\Phi \times H$ = 300 mm × 100 mm were generated and filled with particles according to the coarse contents listed in Figure 2. (2) All the particles were bound by an open cylinder, a bottom disk and a top disk, and then the small cylinders were removed. The initial contact forces and velocities were decreased to 0 for the 5 cycles before the 20000 test cycles, to eliminate the large contact forces between particles in steps (1) and (2), respectively. (3) The specimen was subjected to isotropic compression. Meanwhile, a servo-control mechanism acted on all the walls to ensure the required confining stress of 500 kPa. The assembly was considered to reach equilibration when the ratio of the mean static unbalanced force to the mean contact force was less than 10^{-5} and the tolerance of the stress from the walls was less than 0.5%. The details are shown in Figure 4, where Figure 4a–c corresponds to steps (1)–(3). The friction coefficients between the particle–particle and particle–wall contacts were set to 0 to obtain the densest assemblies [32], which correspond to the compacted aggregates.

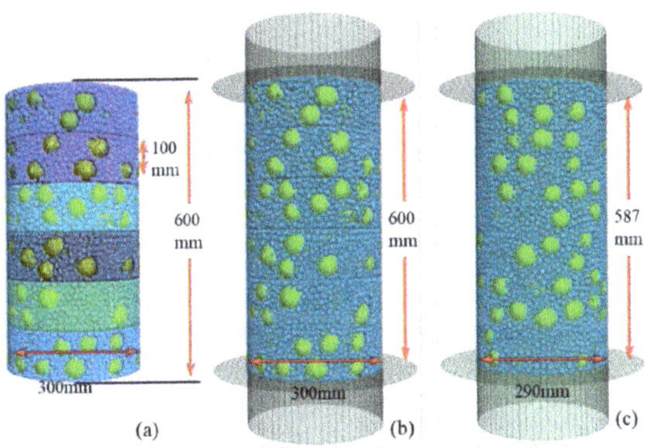

Figure 4. Specimen preparation process: (**a**) generate particles; (**b**) eliminate the large contact forces; (**c**) prepare specimen.

The series of simulations were carried out using a simple linear force-displacement contact law, which resembled the Hertz–Mindlin model, in terms of both the microscopic and macroscopic mechanical behaviors of granular material with spherical and non-spherical particles [25,33]. It has been used in several previous studies [34–36]. For the linear model, the main microscopic parameters are the normal contact stiffness k_n, shear contact stiffness k_s and the coefficient of friction μ_b between particles. The k_n and k_s mainly affect the macroscopic elastic modulus of the particle aggregates and are assigned by the effective modulus E_c, based on the linear contact model in PFC^{3D} ($k_n = \frac{\pi[\min(R_a,R_b)]^2}{R_a+R_b}E_c$), where R_a and R_b are the radii of the spheres in contact. The μ_b mainly affects the shear strength. The selection of these parameters has been verified in [10], and the reliability has also been verified, as presented in Figure 3. Specifically, the effective modulus E^* was 1×10^9 Pa, the friction coefficient μ_b between particles was 0.6, the friction coefficient between the particles and wall was 0, the value of k_n/k_s was 4/3, the damping constant β was 0.7, and the particle density ρ was 2650 kg/m^3.

Three coarse particle contents (25%, 50% and 75%) and six *Md* values (0.0, 0.2, 0.4, 0.6, 0.8 and 1.0) of assemblies were prepared at their densest condition for the generation procedure. Quantitative relationships between mixing degree and strength indices, as well as stress–strain relationships were investigated. In addition, the relationships between mixing degree and particle-scale characteristics, such as coordination number, fabric structure, contact types, contact numbers and forces, were studied. Moreover, the accuracy of the stress–force–fabric based on the branch vector frame was analyzed, and any fabric anisotropies affected by mixing degree were evaluated in the peak and critical states. This detailed analysis of fabric anisotropies allowed us to determine the microscopic mechanisms underlying the dependence of shear strength on mixing degree.

3.2. Simulation Parameters

In conventional triaxial tests, the effective mean stress (p') and deviatoric stress (q) are respectively defined as:

$$p' = (\sigma_1 + \sigma_2 + \sigma_3)/3 \tag{2}$$

$$q = (\sigma_1 - \sigma_3) \tag{3}$$

where σ_1 denotes the axial stress and $\sigma_2 = \sigma_3$ denote the lateral stresses. The axial strain (ε_1) and the volumetric strain (ε_v) can be estimated from the boundary displacements:

$$\varepsilon_1 = (h_i - h_c)/h_i \tag{4}$$

$$\varepsilon_v = (v - v_0)/v_0 \tag{5}$$

where h_i and h_c are the initial height and current height of the specimen, respectively, and v_0 and v are the initial volume and the current volume, respectively. The dilatancy angle (ψ) is defined in triaxial tests as follows:

$$\sin\psi = \frac{d\varepsilon_v/d\varepsilon_1}{2 + d\varepsilon_v/d\varepsilon_1} \tag{6}$$

The internal angle of friction (φ), which reflects the shear strength of granular materials, in the peak or critical state, can be defined as follows:

$$\sin\varphi = \frac{\sigma_1 - \sigma_3}{\sigma_1 + \sigma_3} = \frac{3q/p'}{q/p' + 6} \tag{7}$$

A positive correlation between the angle of internal friction (φ) and the effective stress ratio (q/p') has been established, so it is appropriate to use q/p' to represent the shear strength of the binary geotechnical mixtures.

4. Results and Analyses

4.1. Macroscopic Variation in Strength with Mixing Degree

4.1.1. Shear Strength

This study focused on quantifying the influence of the mixing degree on the macroscopic characteristics, especially the shear strength (q/p'), and identifying their relationship to the micro-characteristics, which cannot be observed in physical experiments. Hence, all of the assemblies were sheared to approximately 40% of ε_1 to explore the macroscopic behaviors at both the peak and critical states.

The relationships between the shear strength ratio (q/p') and the axial strain ε_1 for the various mixing degrees are presented in Figure 5. Since there is no quantitative analysis presented here and there will be a considerable amount of overlap if all 18 curves in this paper are presented, only the results of Md = 0.0 and Md = 1.0 for each tested coarse content are presented here. It can be observed that the q/p' value rapidly rises to the peak value within a strain value (ε_1) of 0.00–0.02, then gradually declines within ε_1 = 0.02–0.15, and eventually approaches the critical state. When ε_1 is greater than 0.30, the curve is more stable. Hence, the average value of q/p' within the axial strain interval [0.30, 0.40] is taken as the critical strength.

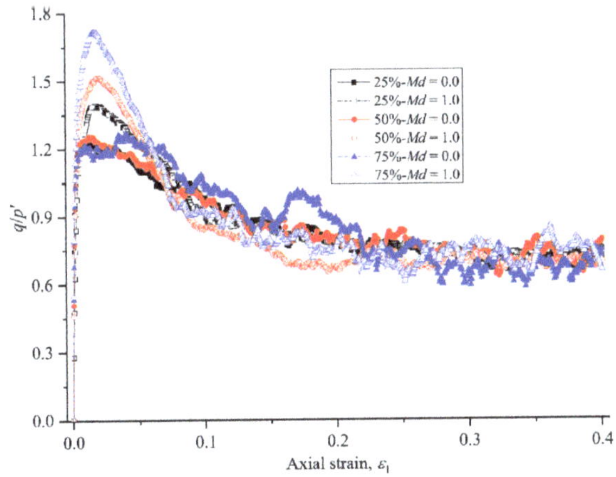

Figure 5. The q/p' versus axial strain ε_1, for specimens with different mixing degrees.

The influence of the mixing degree on the peak shear strength (q/p'_{p}) and the critical shear strength (q/p'_{c}) are as depicted in Figure 6. The q/p'_{p} increases with an increase in mixing degree, and the increasing trend is more obvious with increasing coarse content. However, q/p'_{c} fluctuates slightly within a certain range. The peak shear strength exhibits a monotonic increase as the mixing degree increases, and the critical shear strength was found to be independent of the mixing degree.

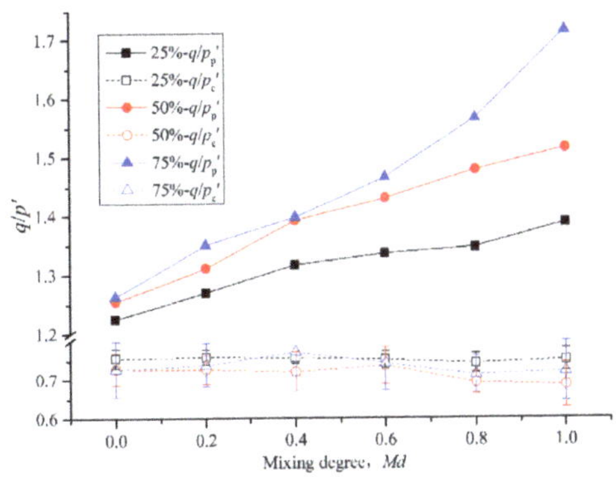

Figure 6. The peak and critical shear strengths as a function of mixing degree.

4.1.2. Volumetric Strain and Dilatancy Angle

As the contact between the binary geotechnical mixture particles is nonplanar, the particles are staggered, which leads to the behavior of particle translation and rotation on the shear plane during the shear process, along with the change in volumetric strain.

Figure 7 shows the evolution of the volumetric strain ε_v with the axial strain ε_1 and presents the relationship between the results of volumetric strain against the axial strain. It can be seen that, at the initial stage of the test, there was no significant difference in strain values for respective specimens until the critical state was attained. This behavior is inconsistent with the volumetric strain behavior of typical dense granular materials [27,37].

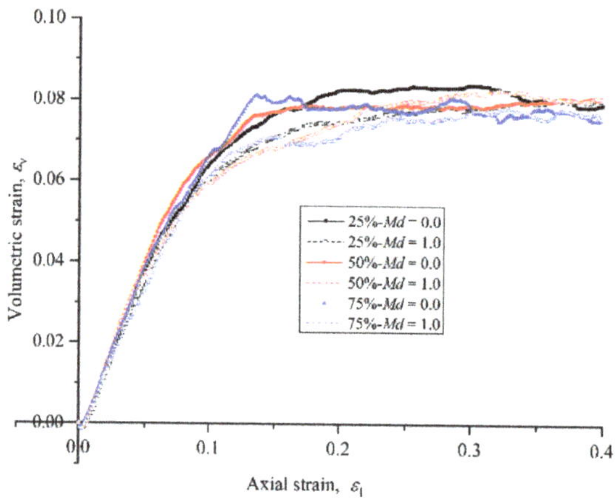

Figure 7. Evolution of the volumetric strain (ε_v) with the axial strain (ε_1).

As it is difficult to distinguish the change in volumetric strain with mixing degree in Figure 7, the peak shrinkage value and critical dilatancy value are shown in Figure 8 in detail. The volumetric strain (shrinkage value) of the specimen in the peak state increases with an increase in mixing degree, while the volumetric strain (dilatancy value) at the critical state is not affected by the change in mixing degree. In fact, both the peak strength and critical strength of soil are closely related to volumetric strain, e.g., the critical point from shrinkage to dilatancy corresponds to the peak shear strength of soil, while the stable dilatancy state corresponds to the critical shear strength [38]. Therefore, it can easily be explained why the variation trend of volumetric strain with mixing degree in the peak or critical state is highly consistent with the variation trend of strength with mixing degree. The volumetric strain has a significant influence on both the peak strength and critical strength, and the quantitative analysis is carried out below.

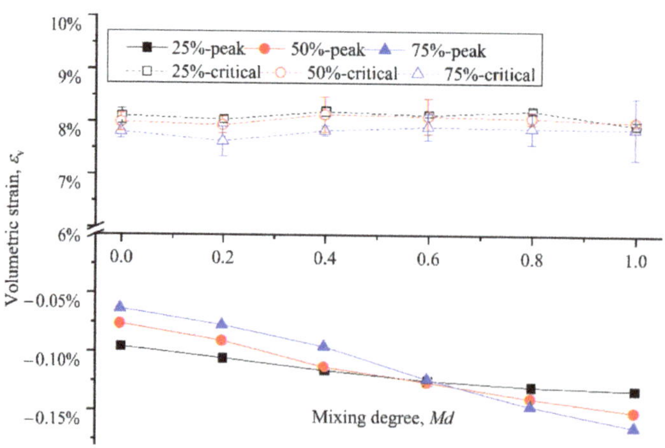

Figure 8. The peak shrinkage and critical dilatancy as a function of *Md*.

Volume expansion can be indicated in terms of the dilatancy angle (ψ). The maximum dilatancy angle is called the peak dilatancy angle ψ_p, which is closely related to the peak shear strength [39]. Figure 9 displays the variation in the peak dilatancy angle with mixing degree. The peak dilatancy angle is not influenced by the mixing degree of the specimen, except that it increases slightly in the specimen with Vf = 75%. This is quite different from the trend of the peak shear strength with the mixing degree. It could be that the variation in the peak shear strength in specimens with different mixing degrees is not directly related to the peak dilatancy angle.

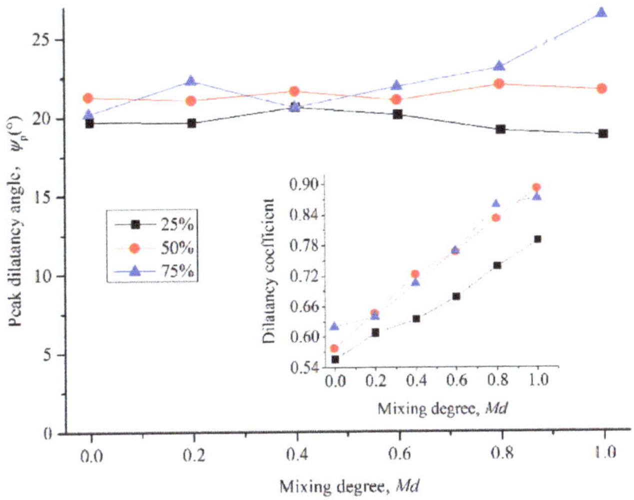

Figure 9. The peak dilatancy angle and dilatancy coefficient as a function of *Md*.

Bolton [38] describes the relationship between the dilatancy angle and shear strength with Equation (8), which is a simplification of an expression presented by Rowe [40]:

$$\varphi_p = \varphi_c + \alpha\psi_p \tag{8}$$

where φ_p and φ_c represent the peak friction angle and critical friction angle, respectively, and α is the dilatancy coefficient, which represents the contribution of dilatancy to the peak shear strength and, to a certain extent, varies with different soil properties. The evolution of the dilatancy coefficient with the mixing degree is shown in Figure 9. For all samples with a given coarse content, the dilatancy coefficient increases significantly with an increase in mixing degree. This implies that the contribution of shrinkage to peak shear strength is more obvious in specimens with higher mixing degrees. It is also proven that the variation in peak shear strength is directly related to the peak dilatancy angle.

4.2. Mesomechanism of the Variation in Strength with Mixing Degree

The macroscopic analysis shows that the coarse content only affects the variation degree of the shear strength with mixing degree. Hence, the subsequent microscopic analysis takes the result of the specimen with a 50% coarse content as the bench mark for the subsequent analysis.

4.2.1. Contact Number

The coordination number is an important parameter for measuring the internal structural characteristics of granular mixtures. The coordination number depends on the type contact between particles. The contact networks in this study were divided into three subnetworks: (i) a coarse–coarse (cc) contact network, (ii) coarse–fine (cf) contact network, and (iii) fine–fine (ff) contact network. The overall coordination number can also be divided into three partial coordination numbers Z_{cc}, Z_{cf}, and Z_{ff}, defined as follows [35,41]:

$$Z_{cc} = 2C_{cc}/N_c \tag{9}$$

$$Z_{cf} = 2C_{cf}/N \tag{10}$$

$$Z_{ff} = 2C_{ff}/N_f \tag{11}$$

where C_m indicates the contact number corresponding to the m contact type (m can be cc, cf and ff). N is the total number of all the particles, N_c is the number of coarse particles and N_f is the number of fine particles.

Hence, the partial coordination number represents the number of different contact types. The coordination is used to quantify the extent to which the different contact types contribute to the shear strength [42]. Figure 10 shows the evolution of the partial coordination numbers and the contribution of each contact type to the shear strength with the mixing degree, where S_{cc}, S_{cf} and S_{ff} represent the contributions of the cc contacts, cf contacts and ff contacts to the shear strength, respectively. It can be clearly observed that, in the peak and critical state, the partial coordination number and the contribution to the shear strength of different contact types follow the same trend. The partial coordination number and the contribution to the shear strength of cc contacts and ff contacts monotonically decrease with an increase in mixing degree, and those of cf contacts monotonically increase with an increase in mixing degree. An intuitive explanation for these results is that as the mixing degree increases, the distribution of coarse and fine particles changes from polarized to uniform. Hence, the contacts between coarse and fine particles become more common, accompanied by the inevitable decrease in the number of contacts between fine particles and between coarse particles. Then, the puzzle is whether the change in the number of contacts affects the shear strength. For example, in the case that the numbers of cc and ff contacts decrease with increasing mixing degree, the increase in peak shear strength with an increase in mixing degree seems to be due to the increased number of cf contacts, while the critical shear strength is independent of these changes. The contribution of different contacts to shear strength varies significantly; this could be attributed to the change in contact numbers.

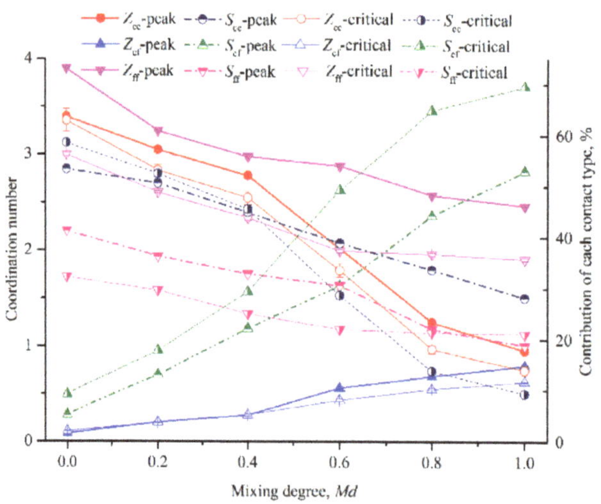

Figure 10. Evolutions of the partial coordination number and the contribution of each contact type to the shear strength with *Md*.

4.2.2. The Fabric Structure

Gong et al. [43] divided the fabric structure of a binary mixture into four types (S1, S2, S3 and S4) according to different coarse contents (i.e., $Vf \leq 30\%$ in S1; $30 < Vf \leq 48\%$ in S2; $48 < Vf \leq 70\%$ in S3; $Vf > 70\%$ in S4). And the contribution of contact type to strength was varied in different structures (i.e., ff > cf > cc in S1; cf > ff > cc in S2; cf > cc > ff in S3; cc > cf > ff in S4). It is worth pointing out that the results in Figure 10 do not strictly satisfy this rule. In other words, all the samples in Figure 10 belong to the S3 structure as a whole, but not all the contributions of contact type to shear strength meet the condition of cf > cc > ff. This is attributed to the mixing degree. It was found that the mixing degree has a direct influence the change of local coarse content, which leads to the differences in local fabric structure from the whole, as shown in Figure 11.

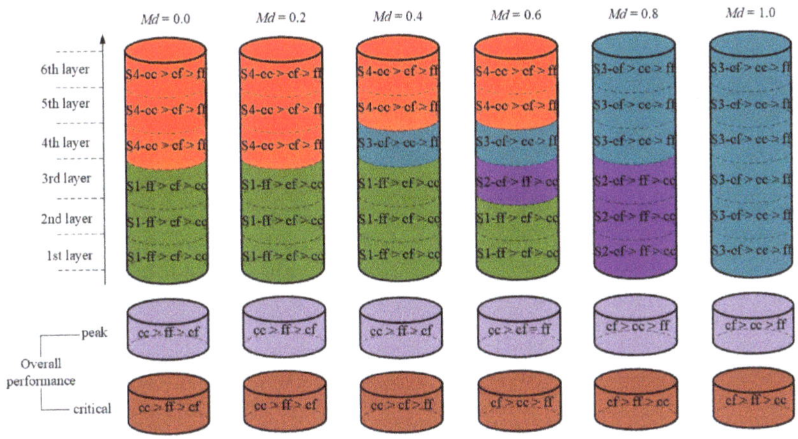

Figure 11. The variation of local fabric structures with depth.

It can be seen from Figure 11, when the mixing degree is low (i.e., $Md = 0.0$ and 0.2), the local fabric structures are only S1 and S4 in the sample. Therefore, in both peak and critical states, cc contact plays a dominant role in contribution to the strength, followed by ff, and finally cf. With the improvement in mixing degree (i.e., $Md = 0.4$ and 0.6), S2 and S3 fabric structures gradually appeared in the middle of the sample, and the contribution of cf contact to strength began to increase in the overall performance. When the samples were uniformly mixed (i.e., $Md = 1.0$), S3 dominated the local fabric structure, and cf contact played a dominant role in the contribution to strength; this is consistent with the previous studies [43–45]. It should be pointed out that the analysis object is the sample with $Vf = 50\%$, and the value is at the junction of the S2 structure and S3 structure. Therefore, in the peak state, the contribution of cc contact to strength is greater than that of ff contact (the overall performance is biased to S3), while in the critical state, the contribution of ff contact to strength is greater than that of cc contact (the overall performance is biased to S2).

In general, the analysis results of Figures 10 and 11 show that with the mixing degree increased, the most intuitive change on the mesoscopic level is that the cf contact number and its contribution to strength increases gradually. However, this change causes different results in peak and critical state; the peak strength gradually increases and the critical strength remained constant, which need further discussion.

4.2.3. Contact Force

In fact, the variation in the contributions of each contact type to the shear strength may be caused by two factors: the contact number and the contact force. When the contact numbers are insufficient to explain the variation in shear strength, the magnitude of the contact force is taken into account. The mean normal contact force $\overline{f}_n$ is defined as the average of the overall normal contact force of the contacts. Similarly, $\overline{f}_{n}$_m represents the mean normal force of m contact type (m can be cc, cf and ff). Table 1 enumerates the contact numbers and the mean normal contact forces $\overline{f}_{n}$_m of various specimens.

Table 1. The contact numbers and the mean normal contact forces $\overline{f}_{n}$ _m of various specimens.

State	Specimen	$\overline{f}_{n}$_cc(N)	$\overline{f}_{n}$_cf(N)	$\overline{f}_{n}$_ff(N)	$\overline{f}_{n}$(N)	C_{cc}	C_{cf}	C_{ff}
peak	$I = 0.0$	2795.68	250.28	117.78	149.95	403	756	36,211
	$I = 0.2$	2796.66	246.22	126.88	160.88	362	2062	33,239
	$I = 0.4$	2664.39	250.87	125.23	162.89	309	3083	30,310
	$I = 0.6$	2936.14	248.44	127.54	164.82	227	5230	27,208
	$I = 0.8$	2833.42	256.10	129.67	161.64	116	6714	25,614
	$I = 1.0$	3272.87	255.50	124.99	169.73	96	7802	20,904
critical	$I = 0.0$	2104.77	326.81	114.41	144.26	398	1173	31,348
	$I = 0.2$	2153.30	312.60	113.70	149.69	343	2099	27,270
	$I = 0.4$	1997.71	302.12	113.09	154.18	312	3006	24,827
	$I = 0.6$	1853.70	290.62	106.88	154.75	210	4627	20,895
	$I = 0.8$	1718.37	274.28	96.98	139.63	97	5863	19,102
	$I = 1.0$	1649.24	245.63	99.41	136.64	76	6657	18,589

It can be observed that, for all the specimens, $\overline{f}_{n}$_cc is the largest, followed by $\overline{f}_{n}$_cf, and $\overline{f}_{n}$_ff is the smallest, which is consistent with previous numerical results [42,46,47]. With an increase in mixing degree, in the peak state, $\overline{f}_{n}$_cc, $\overline{f}_{n}$_cf and $\overline{f}_{n}$_ff all fluctuate within a certain range, while C_{cc} and C_{ff} decrease significantly and C_{cf} increases obviously. Thus, the increase in contribution of cf contacts to peak shear strength is undoubtedly determined to be caused by the increase in the cf contact number. In the critical state, $\overline{f}_{n}$_cc and $\overline{f}_{n}$_ff decrease slightly, C_{cc} and C_{ff} decrease significantly, while $\overline{f}_{n}$_cf decreases and C_{cf} increases significantly with increasing *Md*. At this point, it can be revealed that the effect of the increase in cf contact number on the critical strength is offset by the decrease in the other five factors (e.g., $\overline{f}_{n}$_cc, $\overline{f}_{n}$_cf, $\overline{f}_{n}$_ff, C_{cc} and C_{ff}), especially $\overline{f}_{n}$_cf, resulting in the stability of the critical shear strength. However, such a simple conclusion does not seem sufficient to explain the mechanism of the change in critical shear strength. Therefore, further analysis needs to be carried out on the distribution of contact types.

4.2.4. Distribution of Contacts

Apart from other factors (e.g., shape and size), the most intuitive influence of mixing degree on the specimen is the change in the distribution of coarse and fine particles, resulting in different contact distributions. Based on the results obtained, as presented in Table 1, the cf contact type is the most noteworthy. Therefore, the change in the distribution of cf contacts in specimens with different mixing degrees is shown in Figure 12. In Figure 12, the blurred green and blue spheres represent coarse and fine particles, respectively, while the solid red bars represent cf contacts. It can be seen clearly that the distribution of cf contacts gradually changes from a sporadic distribution at the interface when $Md = 0.0$ to a global distribution at $Md = 1.0$. This means that with the increase in mixing degree, the number of cf contacts in the sample increases and the distribution becomes more uniform. The magnitudes of $\overline{f}_{n}$_cf and $\overline{f}_{n}$ were compared in order to ascertain if the influence of cf contact number on the shear strength is controlled by the contact distribution in the specimen.

The network of contacts that are used to carry forces can be divided into two subnetworks, using strong and weak contacts [48]. The strong contacts can carry larger forces than the mean normal contact force $\overline{f}_{n}$, whereas the weak contacts carry smaller forces than the mean normal contact force $\overline{f}_{n}$. In the shearing of granular materials, strong contacts can work together to transfer force, and the weak contacts provide stability under the forces propagating along the strong contacts. By comparing the values between the mean normal contact forces $\overline{f}_{n}$ and $\overline{f}_{n}$_m, it is not difficult to find that $\overline{f}_{n}$_cc and $\overline{f}_{n}$_cf are much larger than $\overline{f}_n$, while $\overline{f}_n$_ff is less than $\overline{f}_n$. These results indicate that the cc and cf contacts are dominated by strong contacts, whereas the ff contacts are dominated by weak contacts. It is clear that only an increase in C_{cf} can offset the effect of the reduction in $\overline{f}_n$_cc, $\overline{f}_n$_cf, $\overline{f}_n$_ff, C_{cc} and C_{ff} and maintain the stability of the critical shear strength: with the increase in mixing degree, cf contacts gradually spread through the strong contact network from the interface throughout the specimen. At this point, the strong contacts, which can transfer forces, are homogeneously distributed throughout the whole specimen. This further confirms the claim that the increase in peak shear strength is caused by the increase in the cf contacts, which have a similar distribution in the critical state.

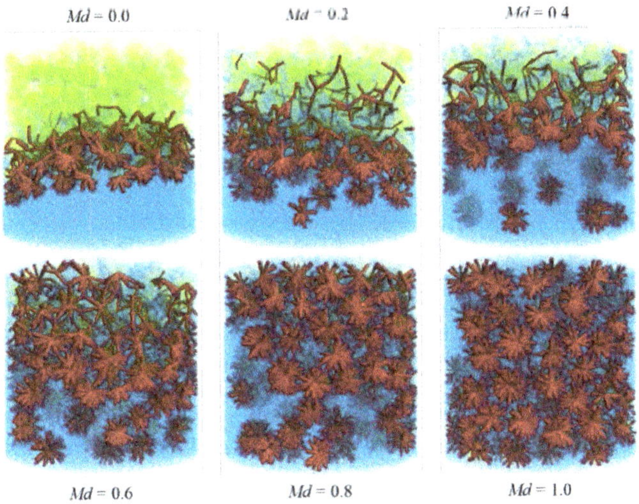

Figure 12. Distribution of cf contacts in the specimens when ε_1 = 0.4.

4.2.5. Force–Fabric Anisotropies

In terms of micromechanics, the force–fabric anisotropies can be regarded as the source of the shear resistance of granular materials [25,49,50]. In order to establish the relationship between macroscopic strength (q/p') and microscopic force–fabric anisotropies of cohesionless granular materials under triaxial condition, the following formula is usually used:

$$q/p' = 0.4a_c + 0.4a_n + 0.6a_t \tag{12}$$

where a_c, a_n, and a_t indicate the normal contact anisotropy, normal contact force anisotropy, and tangential contact force anisotropy, respectively, which can be calculated from the generalized fabric and force tensors, and have been confirmed by previous studies [48,51].

In the three subnetworks mentioned earlier, Equation (12) can also be represented as

$$q/p' = \sum_{m=1}^{k} \xi_m (0.4a_c^m + 0.4a_n^m + 0.6a_t^m) \tag{13}$$

where ξ_m denotes the weighting parameter of the mth subnetwork (i.e., the cc, cf and ff contacts); a_*^m reflects the anisotropy coefficient of contacts in the mth subnetwork and is obtained by restricting the equations of a_* to the mth subnetwork.

Figure 13 shows the variations in anisotropies with Md. It can be seen from Figure 13 that, in both states (the peak state and the critical state), $0.4a_n$ dominates the strength, followed by $0.4a_c$, and then $0.6a_t$. The difference is that, in the peak state, both $0.4a_c$ and $0.6a_t$ decrease slightly with an increase in Md, while $0.4a_n$ increases significantly, which means that the increase in peak strength is mainly caused by the increase in $0.4a_n$. In the critical state, both $0.4a_n$ and $0.6a_t$ increased with the increase in Md, while $0.4a_c$ decreased significantly. This implies that the contributions of these two changes in the strength offset each other and maintain the stability of the critical strength.

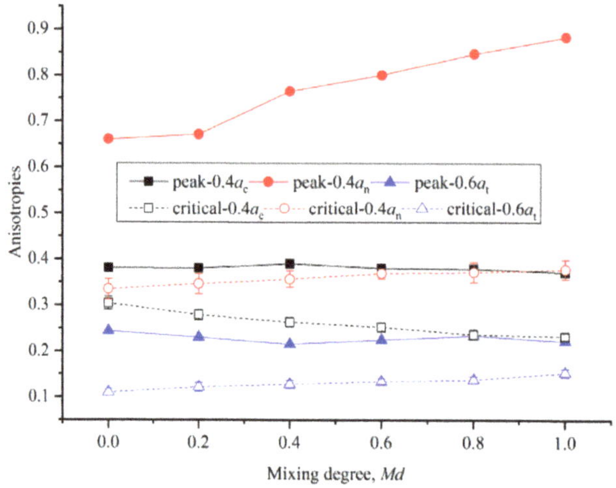

Figure 13. Variations in anisotropies with respect to *Md*.

Considering that the force–fabric anisotropies in the binary geotechnical mixture are determined by the three subnetworks, as quantified in Equation (13), Figure 14a–c shows the evolution of the anisotropies for the various contact types as a function of *Md*. In the cc contact network, in the peak state, $0.4a_c$, $0.4a_n$ and $0.6a_t$ all decrease significantly with an increase in *Md*, while at the critical state, $0.4a_c$ decreases slightly, $0.4a_n$ is relatively stable, and $0.6a_t$ increases slightly. In the cf contact network, $0.4a_c$, $0.4a_n$ and $0.6a_t$ are all significantly increased with the increase in *Md* in the peak state, while $0.4a_n$ is significantly increased, $0.4a_c$ shows a downward trend, and $0.6a_t$ is basically stable in the critical state. In the ff contact network, all the force–fabric anisotropy coefficients show a slight downward trend regardless of the peak or critical state.

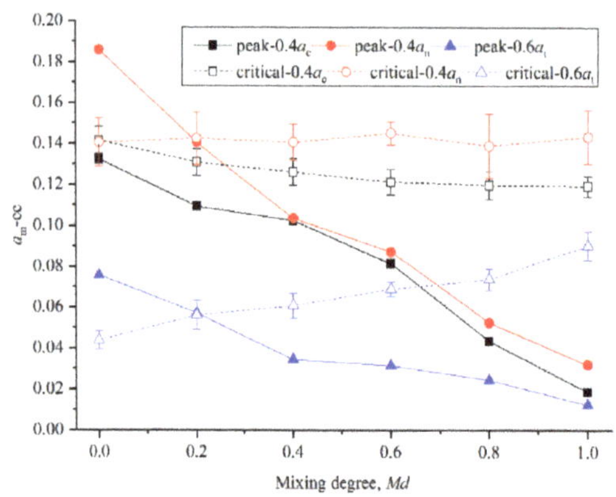

(a)

Figure 14. *Cont.*

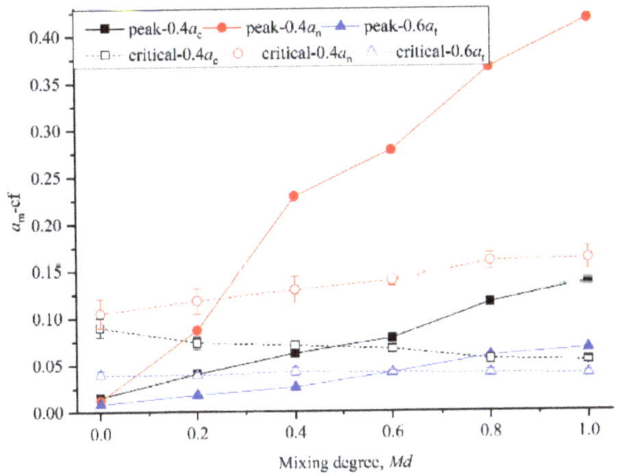

(b)

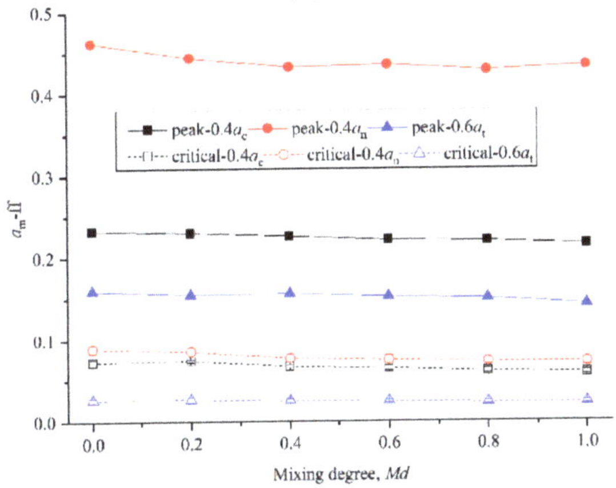

(c)

Figure 14. Evolutions of anisotropies of different contact types as a function of *Md*: (**a**) cc contact; (**b**) cf contact; (**c**) ff contact.

It can be seen that, in the peak state on the premise, the force–fabric anisotropies in the cc and ff contact networks decrease, and increase significantly in the cf contact network, which results in an increase in peak strength. In the critical state, although the force–fabric anisotropies of the sub-contact network change to some extent, the trend of these changes is relatively gentle, and plays a role in mutual cancellation, resulting in the stability of the critical strength. This result confirms the previous conclusion that the change in strength is mainly caused by the changes in the cf contact number and contact force.

4.3. *Influence of Coarse Content on the Increased Range of Peak Shear Strength*

As shown in Figure 6, under the same mixing degree, the peak shear strength increases monotonically with increasing coarse content (which is increased from 25% to 75%). This is consistent with the results of previous studies on homogeneous specimens; that is, the maximum peak shear strength appears at a coarse content of approximately 75% [52,53]. However, another noteworthy phenomenon is that the increased range of peak shear strength with mixing degree is not consistent under different coarse contents. This means that the mixing degree has different influence levels on the peak shear strength of binary geotechnical mixtures with different coarse contents. Figure 15 shows the curve of normalized peak shear strength changing with mixing degree, where the values of the homogeneous specimens (*Md* = 1.0) are taken as the control. In order to make the conclusion more convincing, the results of samples with 37.5% and 62.5% coarse contents are added here. It can be clearly seen that the higher the coarse content, the more the peak strength is affected by the mixing degree. Specifically, as the coarse and fine particles in the specimens change from homogeneous mixing (*Md* = 1.0) to complete separation (*Md* = 0.0), the peak shear strength of the specimens decreases by 11.7% at 25% coarse content, while it decreases by 17.2% at 50% coarse content and decreases by 26.4% at 75% coarse content.

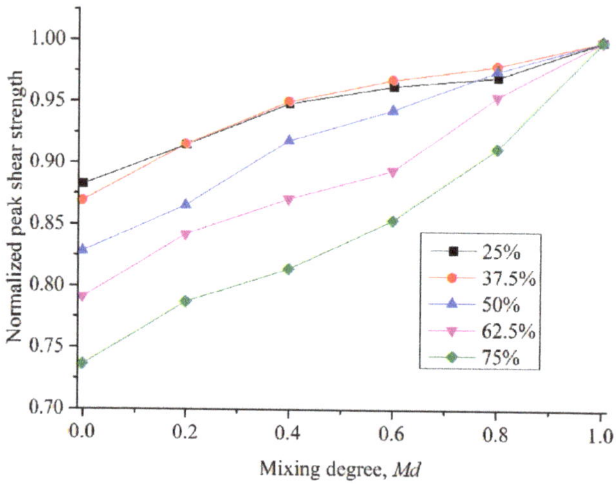

Figure 15. Normalized peak shear strength as a function of mixing degree.

Some previous studies have shown that the binary geotechnical mixture exhibits many "optimal properties" when the coarse content is approximately 75%, and all of them show monotonically increasing properties with coarse content range of 25–75%. For example, the minimum void ratio results in the maximum peak shear strength [52]; in this case, the cf contact number and mean normal force also reach the maximum, while the sliding fraction of the cf contact is the minimum [54]. The cf contact is proven to be the main factor affecting the peak shear strength in this paper. On the other hand, for granular materials, the shear strength mainly depends on the interlocking effect of particle accumulation [55]. Meanwhile, a previous study shows that the average degree of interlocking of a particle increases with increasing coarse content [43]. The above conclusions show that when the coarse content is 75% and the specimen is homogeneously mixed (*Md* = 1.0), all the indices correspond to the peak shear strength. The mixing degree changes the local coarse content of the specimen. Regardless of the coarse content, the homogeneous specimen gradually exhibits complete segregation with further mixing degrees; that is, coarse and fine particles occupy separate parts of the specimen, as shown in Figure 2. Hence, there is no doubt that the closer the homogeneous specimen is to the "optimal state", the more obvious the loss in strength when the mixing degree gradually changes to 0.0. That is, the closer the coarse content is to 75%, the more the peak shear strength is affected by the mixing degree.

According to the definition of Scholtes et al. [56], the degree of interlocking of a particle *p* can be estimated through the assessment of its mean internal moment m^p, and Gong et al [43] further stated that the average degree of interlocking (*ADI*) of a particle in the sample can be defined as: $ADI = \sum_{p \in N} m^p / N$. Therefore, the evolution of *ADI* with respect to *Vf* is illustrated in Figure 16, which can be used to quantify the interlocking effect of samples with different coarse contents. It was clearly observed that the evolutions of *ADI* in the samples of *Md* = 0 and *Md* = 1 with *Vf* are roughly consistent. Specifically, the *ADI* value increases with the increase in *Vf*, and the increased range is wide continuously. This is in agreement with the conclusion of Gong et al [43]. It can be seen that the difference of *ADI* between *Md* = 0 and *Md* = 1 increases with the increase in *Vf*, and the increased range increased sharply with the increase in *Vf*. Thus, the higher the *Vf* the more the interlocking effect of the sample is affected by the mixing degree. This result, as an example, can be used to explain why the closer the coarse content is to 75%, the more the peak shear strength is affected by the mixing degree.

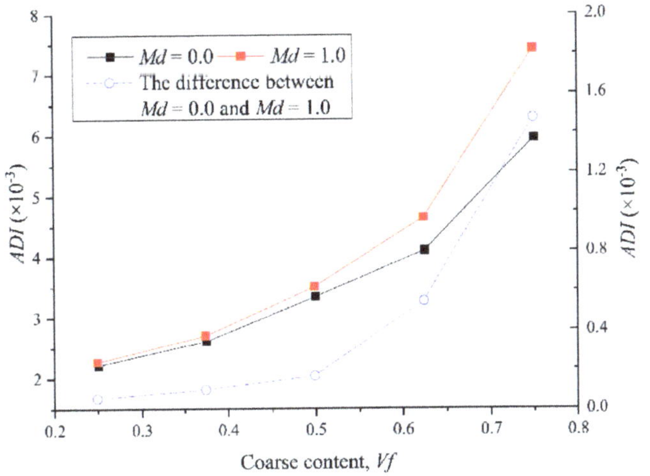

Figure 16. Evolution of the average degree of interlocking of a particle in the sample *ADI* with respect to *Vf*.

5. Conclusions

In this paper, the effect of the mixing degree on both the peak and critical shear strengths of binary geotechnical mixtures was quantitatively analyzed through verified DEM simulations. The results can guide the construction of marine geotechnical engineering, such as more accurate calculation of foundation stability and screw anchor bearing capacity. The samples with three coarse contents (ranging from 25% to 75%) and six mixing degrees (ranging from 0.0 to 1.0) were prepared to create specimens of binary geotechnical mixtures. Both the macroscopic trends of the variations in strength and the meso-mechanisms of the variations in strength were meticulously examined. The main findings are summarized as follows:

(1) The peak shear strength increases with increasing mixing degree, while the critical shear strength is not affected. This trend is consistent with that of volumetric strain with increasing mixing degree, and the variation in peak shear strength is directly related to the peak dilatancy angle.

(2) The variations in both the peak and critical shear strengths are mainly affected by the cf contact type. Specifically, when the mixing degree increases, the mean normal force of cf contacts does not change, while the increase in the number of cf contacts and their global distribution contribute to the increase in peak shear strength. The increase in the cf contact number and the changes in the distribution of cf contacts from local to global offset the influence of the reduction in the mean normal forces of the cc, cf and ff contacts and the cc and ff contact numbers, which results in stable critical shear strength.

(3) The force–fabric anisotropies of the whole contact network and three subnetworks are evaluated. The force–fabric anisotropies of the whole contact network explain the microscopic mechanism of the strength. The increase in peak strength is mainly caused by the increase in $0.4a_n$. And the increase in $0.4a_n$ and $0.6a_t$, with a decrease

in $0.4a_c$, offsets each other, which maintain the stability of the critical strength. The meticulous analysis of the force–fabric anisotropies of three subnetworks affords a way to explain the micro-mechanisms that result in the dependency of the contributions of different contact types to the strength on the mixing degree.

(4) The range of coarse contents conformed to the definition of binary geotechnical mixture and the influence of the mixing degree on the peak shear strength increases with increasing coarse content. This means that with a higher coarse content, a greater influence of the mixing degree on the shear strength performance of binary geotechnical mixture should be considered.

Author Contributions: Conceptualization, S.L. and Z.L.; methodology, M.A.; validation, Y.N. and W.H.; resources, Z.L.; data curation, S.L.; writing—original draft preparation, S.L.; writing—review and editing, Y.N.; supervision, W.H.; project administration, Z.L.; funding acquisition, J.Z. All authors have read and agreed to the published version of the manuscript.

Funding: The study was funded by the National Natural Science Foundation of China, (Nos. 51809292, 52078500 and 52178332) and the Beijing Municipal Science and Technology Project: Research and Application of Design and Construction Technology of Railway Engineering Traveling the Rift Valley, China (No. Z181100003918005).

Institutional Review Board Statement: Not applicable.

Informed Consent Statement: Not applicable.

Data Availability Statement: Not applicable.

Conflicts of Interest: The authors declare no conflict of interest.

References

1. Huang, Y.; Jin, P. Impact of human interventions on coastal and marine geological hazards: A review. *Bull. Eng. Geol. Environ.* **2018**, *77*, 1081–1090. [CrossRef]
2. Fateh, A.M.A.; Eslami, A.; Fahimifar, A. Direct CPT and CPTu methods for determining bearing capacity of helical piles. *Mar. Georesources Geotechnol.* **2016**, *35*, 193–207. [CrossRef]
3. Huang, Y.; Bao, Y.; Zhang, M.; Liu, C.; Lu, P. Analysis of the mechanism of seabed liquefaction induced by waves and related seabed protection. *Nat. Hazards* **2015**, *79*, 1399–1408. [CrossRef]
4. Zhang, M.; Huang, Y.; Bao, Y. The mechanism of shallow submarine landslides triggered by storm surge. *Nat. Hazards* **2016**, *81*, 1373–1383. [CrossRef]
5. Bondevik, S.; Mangerud, J.; Dawson, S.; Dawson, A.; Lohne, Ø. Evidence for three North Sea tsunamis at the Shetland Islands between 8000 and 1500 years ago. *Quat. Sci. Rev.* **2005**, *24*, 1757–1775. [CrossRef]
6. Lee, C.-W.; Kim, Y.-S.; Yoon, Y.-S.; Chang, D.-S.; Choo, Y.W.; Jung, M.-C. Stability Study of a Tide Embankment Subjected to Sea Level Variations Using Centrifugal Model Tests. *Mar. Georesources Geotechnol.* **2015**, *33*, 376–390. [CrossRef]
7. Azema, E.; Preechawuttipong, I.; Radjai, F. Binary mixtures of disks and elongated particles: Texture and mechanical properties. *Phys. Rev. E* **2016**, *94*, 042901. [CrossRef]
8. Molinero Guerra, A.; Mokni, N.; Cui, Y.-J.; Delage, P.; Tang, A.M.; Aimedieu, P.; Bernier, F.; Bornert, M. Impact of initial structural heterogeneity on long-term swelling behavior of MX80 bentonite pellet/powder mixtures. *Can. Geotech. J.* **2020**, *57*, 1404–1416. [CrossRef]
9. Dong, H.; Peng, B.; Gao, Q.-F.; Hu, Y.; Jiang, X. Study of hidden factors affecting the mechanical behavior of soil–rock mixtures based on abstraction idea. *Acta Geotech.* **2021**, *16*, 595–611. [CrossRef]
10. Liu, S.; Hu, W.; Gong, J.; Nie, Z. An improved index of mixing degree and its effect on the strength of binary geotechnical mixtures. *Granul. Matter* **2021**, *24*, 1–19. [CrossRef]
11. Cao, X.; Zhu, Y.; Gong, J. Effect of the intermediate principal stress on the mechanical responses of binary granular mixtures with different fines contents. *Granul. Matter* **2021**, *23*, 1–19. [CrossRef]
12. Rahman, S.; Khattak, M.J.; Adhikari, B.; Adhikari, S. Discrete element modeling of bonded soil mixtures under uniaxial compression and indirect tension test. *Transp. Geotech.* **2021**, *26*, 100438. [CrossRef]
13. Lacey, P.M.C. Developments in the theory of particle mixing. *J. Appl. Chem.* **1954**, *4*, 257–268. [CrossRef]
14. Windows-Yule, C.R.K.; Scheper, B.J.; Horn, A.J.; Hainsworth, N.; Saunders, J.; Parker, D.J.; Thornton, A.R. Understanding and exploiting competing segregation mechanisms in horizontally rotated granular media. *New J. Phys.* **2016**, *18*, 023013. [CrossRef]
15. Cho, M.; Dutta, P.; Shim, J. A non-sampling mixing index for multicomponent mixtures. *Powder Technol.* **2017**, *319*, 434–444. [CrossRef]

16. Siraj, M.S.; Radl, S.; Glasser, B.J.; Khinast, J.G. Effect of blade angle and particle size on powder mixing performance in a rectangular box. *Powder Technol.* **2011**, *211*, 100–113. [CrossRef]
17. Yan, Z.; Wilkinson, S.K.; Stitt, E.H.; Marigo, M. Investigating mixing and segregation using discrete element modelling (DEM) in the Freeman FT4 rheometer. *Int. J. Pharm.* **2016**, *513*, 38–48. [CrossRef]
18. Konan, N.A.; Huckaby, E.D. Kinetic theory-based numerical modeling and analysis of bi-disperse segregated mixture fluidized bed. *Powder Technol.* **2017**, *319*, 71–91. [CrossRef]
19. Dai, B.B.; Yang, J.; Liu, F.T.; Gu, X.Q.; Lin, K.R. A new index to characterize the segregation of binary mixture. *Powder Technol.* **2020**, *363*, 611–620. [CrossRef]
20. Medley, E. Systematic characterization of melange bimrocks and other chaotic soil_rock mixtures. *Felsbau* **1999**, *17*, 152–162.
21. Zhao, H.F.; Zhang, L.M. Effect of coarse content on shear behavior of unsaturated coarse granular soils. *Can. Geotech. J.* **2014**, *51*, 1371–1383. [CrossRef]
22. Yan, X.; Zhan, W.; Hu, Z.; Yu, Y.; Xiao, D.; de Oliveira Correia, J.A.F. Experimental Study on the Effect of Compaction Work and Defect on the Strength of Soil-Rock Mixture Subgrade. *Adv. Mater. Sci. Eng.* **2021**, *2021*, 1–13. [CrossRef]
23. Belheine, N.; Plassiard, J.P.; Donzé, F.V.; Darve, F.; Seridi, A. Numerical simulation of drained triaxial test using 3D discrete element modeling. *Comput. Geotech.* **2009**, *36*, 320–331. [CrossRef]
24. Evans, T.M.; Valdes, J.R. The microstructure of particulate mixtures in one-dimensional compression: Numerical studies. *Granul. Matter* **2011**, *13*, 657–669. [CrossRef]
25. Zhao, S.; Matthew, E.T.; Zhou, X. Shear-induced anisotropy of granular materials with rolling resistance and particle shape effects. *Int. J. Solids Struct.* **2018**, *150*, 268–281. [CrossRef]
26. Abedi, S.; Mirghasemi, A.A. Particle shape consideration in numerical simulation of assemblies of irregularly shaped particles. *Particuology* **2011**, *9*, 387–397. [CrossRef]
27. Gong, J.; Liu, J. Effect of aspect ratio on triaxial compression of multi-sphere ellipsoid assemblies simulated using a discrete element method. *Particuology* **2017**, *32*, 49–62. [CrossRef]
28. Nie, Z.; Zhu, Y.; Wang, X.; Gong, J. Investigating the effects of Fourier-based particle shape on the shear behaviors of rockfill material via DEM. *Granul. Matter* **2019**, *21*, 1–15. [CrossRef]
29. Yang, Y.; Wang, J.F.; Cheng, Y.M. Quantified evaluation of particle shape effects from micro-to-macro scales for non-convex grains. *Particuology* **2016**, *25*, 23–35. [CrossRef]
30. Xu, W.J.; Xu, Q.; Hu, R.L. Study on the shear strength of soil–rock mixture by large scale direct shear test. *Int. J. Rock Mech. Min. Sci.* **2011**, *48*, 1235–1247.
31. Zhou, W.; Xu, K.; Ma, G.; Yang, L.; Chang, X. Effects of particle size ratio on the macro- and microscopic behaviors of binary mixtures at the maximum packing efficiency state. *Granul. Matter* **2016**, *18*, 1–13. [CrossRef]
32. Abbireddy, C.O.R.; Clayton, C.R.I. Varying initial void ratios for DEM simulations. *Géotechnique* **2010**, *60*, 497–502. [CrossRef]
33. Zhao, S.; Evans, T.; Zhou, X. Effects of curvature-related DEM contact model on the macro-and micro-mechanical behaviours of granular soils. *Géotechnique* **2018**, *68*, 1085–1098. [CrossRef]
34. Nie, Z.; Liu, S.; Hu, W.; Gong, J. Effect of local non-convexity on the critical shear strength of granular materials determined via the discrete element method. *Particuology* **2020**, *52*, 105–112. [CrossRef]
35. Minh, N.H.; Cheng, Y.P. A DEM investigation of the effect of particle-size distribution on one-dimensional compression. *Géotechnique* **2013**, *63*, 44–53. [CrossRef]
36. Gu, X.; Zhang, J.; Huang, X. DEM analysis of monotonic and cyclic behaviors of sand based on critical state soil mechanics framework. *Comput. Geotech.* **2020**, *128*, 103787. [CrossRef]
37. Zhu, Y.; Nie, Z.; Gong, J. Influence of the rolling-resistance-based shape of coarse particles on the shear responses of granular mixtures. *Particuology* **2020**, *52*, 67–82. [CrossRef]
38. Bolton, M.D. The strength and dilatancy of sands. *Géotechnique* **1986**, *35*, 65–78. [CrossRef]
39. Simoni, A.; Houlsby, G.T. The Direct Shear Strength and Dilatancy of Sand-gravel Mixtures. *Geotech. Geol. Eng.* **2006**, *24*, 523–549. [CrossRef]
40. Rowe, P.W. The relation between the shear strength of sands in triaxial compression, plane strain and direct. *Geotech.* **1969**, *19*, 75–86. [CrossRef]
41. Thornton, C. Numerical simulations of deviatoric shear deformation of granular media. *Géotechnique* **2000**, *50*, 43–53. [CrossRef]
42. Minh, N.H.; Cheng, Y.P.; Thornton, C. Strong force networks in granular mixtures. *Granul. Matter* **2014**, *16*, 69–78. [CrossRef]
43. Gong, J.; Liu, J.; Cui, L. Shear behaviors of granular mixtures of gravel-shaped coarse and spherical fine particles investigated via discrete element method. *Powder Technol.* **2019**, *353*, 178–194. [CrossRef]
44. De Frias Lopez, R.; Silfwerbrand, J.; Jelagin, D.; Birgisson, B. Force transmission and soil fabric of binary granular mixtures. *Géotechnique* **2016**, *66*, 578–583. [CrossRef]
45. Vallejo, L.E. Interpretation of the limits in shear strength in binary granular mixtures. *Can. Geotech. J.* **2001**, *38*, 1097–1104. [CrossRef]
46. Voivret, C.; Radjai, F.; Delenne, J.Y.; El Youssoufi, M.S. Multiscale force networks in highly polydisperse granular media. *Phys. Rev. Lett.* **2009**, *102*, 178001. [CrossRef] [PubMed]
47. Gong, J.; Liu, J. Mechanical transitional behavior of binary mixtures via DEM: Effect of differences in contact-type friction coefficients. *Comput. Geotech.* **2017**, *85*, 1–14. [CrossRef]

48. Sufian, A.; Russell, A.R.; Whittle, A.J. Anisotropy of contact networks in granular media and its influence on mobilised internal friction. *Géotechnique* **2017**, *67*, 1–14. [CrossRef]
49. Ouadfel, H.; Rothenburg, L. 'Stress-force-fabric' relationship for assemblies of ellipsoids. *Mech. Mater.* **2001**, *33*, 201–221. [CrossRef]
50. Guo, N.; Zhao, J. The signature of shear-induced anisotropy in granular media. *Comput. Geotech.* **2013**, *47*, 1–15. [CrossRef]
51. Markauskas, D.; Kačianauskas, R.; Džiugys, A.; Navakas, R. Investigation of adequacy of multi-sphere approximation of elliptical particles for DEM simulations. *Granul. Matter* **2009**, *12*, 107–123. [CrossRef]
52. Zhu, Y.; Nie, Z.; Gong, J.; Zou, J.; Zhao, L.; Li, L. An analysis of the effects of the size ratio and fines content on the shear behaviors of binary mixtures using DEM. *Comput. Geotech.* **2020**, *118*, 103353. [CrossRef]
53. Zhu, Y.; Gong, J.; Nie, Z. Numerical investigation of the elastic properties of binary mixtures as a function of the size ratio and fines content. *Int. J. Geomech.* **2020**, *20*, 04020155. [CrossRef]
54. Zhou, W.; Wu, W.; Ma, G.; Ng, T.-T.; Chang, X. Undrained behavior of binary granular mixtures with different fines contents. *Powder Technol.* **2018**, *340*, 139–153. [CrossRef]
55. Yao, Y.; Li, J.; Ni, J.; Liang, C.; Zhang, A. Effects of gravel content and shape on shear behaviour of soil-rock mixture: Experiment and DEM modelling. *Comput. Geotech.* **2022**, *141*, 104476. [CrossRef]
56. Scholtès, L.; Hicher, P.-Y.; Sibille, L. Multiscale approaches to describe mechanical responses induced by particle removal in granular materials. *C. R. Mécanique* **2010**, *338*, 627–638. [CrossRef]

sustainability

MDPI

Article

Sensitivity Analysis of Factors Affecting the Bearing Capacity of Suction Bucket Foundation in Soft Clay

Bin Wang [1,2], Ming-Hui Yuan [3], Liang Li [3], Chang-Feng Yuan [3,*], Ying Li [4] and Kan-Min Shen [2]

1 Key Laboratory of Far-Shore Wind Power Technology of Zhejiang Province, Hangzhou 311122, China
2 Powerchina Huadong Engineering Corporation Limited, Hangzhou 311122, China
3 School of Civil Engineering, Qingdao University of Technology, Qingdao 266520, China
4 Chinese–German Institute of Engineering, Zhejiang University of Science and Technology, Hangzhou 310023, China
* Correspondence: yuanchangfeng@qut.edu.cn; Tel.: +86-131-7323-0827

Abstract: A suction bucket is the foundation for the development of offshore wind power technology in the deep sea, and its stability is crucial to the superstructure of the wind power generation system. Combined with soft clay soil strata along the Chinese coast, the bearing capacity of suction bucket foundations was studied using a numerical model. Sensitivity factors such as soil strength with random space distribution, dimensions of foundation, wind and wave loads in different directions, and cycle times were considered. The results show that the normalized foundation bearing capacity coefficient increases with the increase of the foundation length–diameter ratio. When the foundation length–diameter ratio is less than 1.0, the foundation bearing capacity coefficient is more sensitive to the soil non-uniformity coefficient than the length–diameter ratio. When the length–diameter ratio of the suction bucket is large enough, the influence of the soil non-uniformity coefficient on the bearing capacity of the suction bucket foundation gradually diminishes. When the direction of wind and wave loads is 15°, the bearing capacity of the suction bucket foundation is the weakest. Under the cyclic loads, the shallow soil strength weakens faster in the initial stage and the attenuation rate of soil strength slows down in the latter stage.

Keywords: soft clay; offshore wind power; suction bucket; bearing capacity; sensitivity

1. Introduction

Offshore wind power has increasingly been a promising alternative to traditional energy. It is one of the most critical issues that the proper selection of foundation type for the construction of offshore wind power. To date, due to the advantages of fast installation and low cost, the monopile foundation [1–3] is one of the most widely used foundations where the water depth is less than 30 m. However, the difficulty and cost of monopile installation will rapidly increase as the depth of the sea increases, which is not conducive to developing deep-sea wind power. Hence, new types of foundations are constantly developed for the deep sea [4,5]. For instance, the multi-pod foundation is suitable for water depths of 30–50 m, but it is difficult to transport and too complicated to install; the jacket foundation is suitable for water depths between 30–60 m, but it has a long manufacturing and installation cycle as well as a high cost that decreases rapidly with the deepening of water depth; the floating foundation is suitable for water depths above 50 m, but it is challenging to design and has the potential to tilt, shake, and overturn. The suction bucket in particular (Figure 1) is a novel deep-sea foundation suitable for water depths above 30 m [6–10]. Compared with the above foundation types, it has the advantages of convenient installation, lower cost, and recyclability.

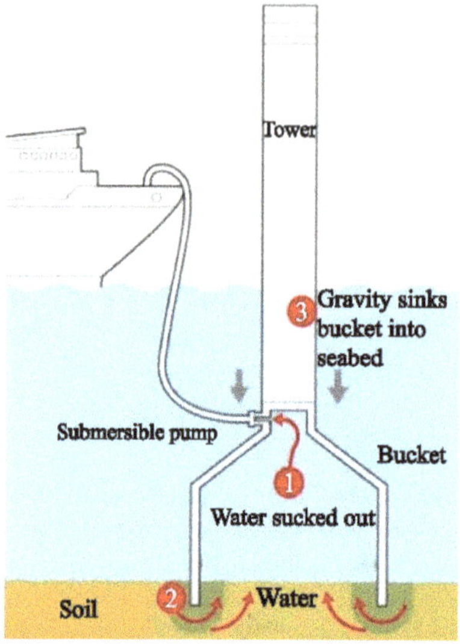

Figure 1. Schematic diagram of suction bucket foundation.

As a new type of foundation for offshore wind power, the suction bucket is primarily a large-diameter barrel with an open bottom and a closed top, which is an alternative form of single pile foundation. The suction bucket has the advantages of low cost, simple construction process, integrated construction, fast installation, and secondary use, which can be better applied to a deep seabed foundation. Meanwhile, the suction bucket foundation [11], which has been effectively developed in recent years, can better resist the large horizontal load and bending moment of wind turbines (Figure 2).

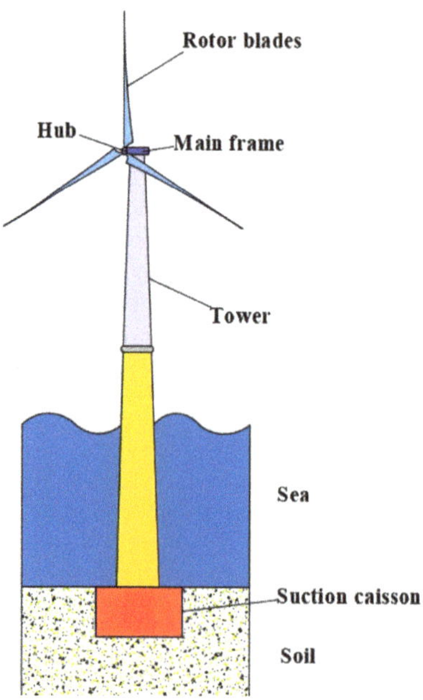

Figure 2. Suction bucket foundation in a wind power generation system.

The study on the bearing capacity characteristics of the suction bucket has attracted much attention. In sand and saturated soft clay, a simple method for estimating the horizontal ultimate bearing capacity of suction bucket foundations has been proposed [12,13]. However, the movement of the foundation rotation center and the complex stress distribution in practical engineering have not been taken into account by this method. Therefore, the calculated bearing capacity was different from the practical engineering. The effect of several swipe loading schemes on the bearing capacity of suction bucket foundations in saturated soft clay and sandy soil has been studied by some scholars. It is worth mentioning that Wu K. and Zhang Y.K.'s results [14,15] are only applicable to foundations in the shallow sea with a small length–diameter ratio. The failure envelope surface of the deep-sea foundations with a large length–diameter ratio needs further study. The limitation of this method is that the actual loads were simplified to the vertical load (V), horizontal load (H), and bending moment (M), which were acted on the midpoint of the top surface of the suction bucket without the horizontal cyclic load. To solve this issue, Wang J.H. [16] studied the effect of vertical loads on the horizontal cyclic bearing capacity of the suction bucket and simulated the suction bucket foundation in soft clay under various vertical and horizontal cyclic loads. According to the results, the vertical load affects the cyclic load amplitude and the number of cycles that lead to the instability and failure of the foundation. The larger the vertical load, the smaller the cyclic load amplitude when the foundation is damaged with the same number of cycles. Therefore, in practical engineering, the vertical load effect caused by various upper fan specifications should be considered and the horizontal bearing capacity should be checked.

According to the Code for Design of Wind Turbine Foundations for Offshore Wind Power Projects (NB/T 10105-2018) [17], the service life of wind turbines should not be less than 25 years. During its 25-year service life, the suction bucket foundation not only bears the long-term effects of its weight and the vertical load transmitted by the superstructure, but also frequently bears the combined effects of horizontal cyclic loads and bending moments caused by wave, ocean currents, wind, etc. When utilized as an offshore wind turbine foundation, it frequently bears larger horizontal loads than vertical loads so that the horizontal load becomes the main controlling load for design. Based on previous engineering experience [18,19], continuous horizontal cyclic loading on the foundation in soft clay will lead to a decrease in soil strength and stiffness, ultimately leading to the destruction of suction bucket foundations before reaching the designed horizontal load. Therefore, it is not entirely reliable to determine the stability of the structure only by relying on the ultimate bearing capacity in the static state. To ensure the accuracy of stability analysis, it is necessary to design a numerical method to accurately predict the ultimate bearing capacity of the suction bucket under horizontal cyclic loads. In this paper, the length–diameter ratio of the suction bucket, soil heterogeneity, the angle between wind and wave loads, and the number of cycles were considered in order to study the bearing characteristics of the suction bucket foundation in saturated soft clay. It is instructive to the analysis of bearing capacity stability in practical engineering.

2. Establishment the Numerical Model

2.1. Overview of the Model

2.1.1. Parameters of the Model

A three-dimensional finite element model of interaction between a single suction bucket and soft clay was established for offshore wind power. The truncated soil model has a radial diameter of 10D (D is the diameter of a suction bucket) and a depth of 6L (L is the length of a suction bucket) [9,14] to ignore boundary effects (Figure 3a). The dimensions of the soil model were verified to be correct by numerical simulation. To improve the computing efficiency, a half model of the bucket–soil system was used for the finite element analysis as the numerical model is symmetrical. The boundary conditions of the soil are as follows: the fixed constraint at the bottom, the lateral radial displacement in the directions of X and Y at the constrained flank of half-cylinder, and the displacement in the direction of X at the constrained symmetric cross section. A schematic diagram of the model forces is shown in Figure 3b. F_t represents the force transmitted by the superstructure, p represents the pressure difference between the pressure inside the bucket and the atmospheric pressure, $F_{fric\text{-}in}$ and $F_{fric\text{-}out}$ represent the frictional forces generated by the inside and outside of the bucket wall against the soil, and F_Q represents the shear force.

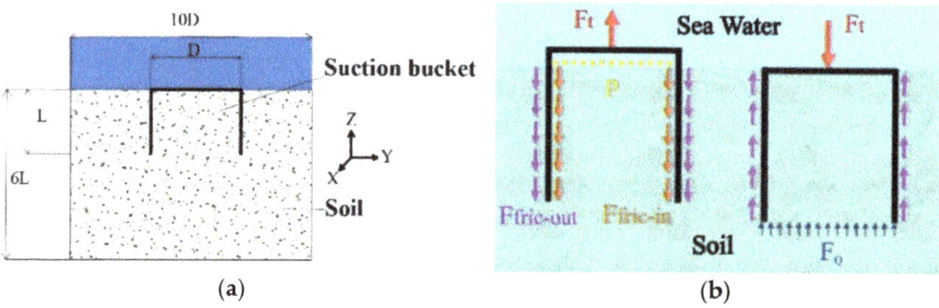

Figure 3. The dimensions and forces of the model. (**a**) Dimensions. (**b**) Forces.

The linear elastic constitutive model with elastic modulus was used to simulate the suction bucket, as shown in Table 1. The Mohr–Coulomb elastoplastic constitutive model [13,14,16] is used to simulate the soil, as shown in Table 2. It is assumed that the soil interacts with the suction bucket under completely undrained conditions [2,9,14].

Table 1. Parameters of suction bucket.

Parameters	Elasticity Modulus/GPa	Poisson's Ratio	Effective Weight/(KN·m^{-3})
Bucket	210	0.125	23

Table 2. Parameters of soil.

Parameters	Elasticity Modulus	Poisson's Ratio	Effective Weight/(KN·m^{-3})	Cohesive Force/KPa	Internal Friction Angle/(°)	Dilatancy Angle/(°)
Soil	$500S_u$	0.49	8	20	15	0.1

2.1.2. Mesh

Both the suction bucket and the soil use an integral three-dimensional solid element (C3D8R) with an 8-node hexahedral linear reduction grid (Figure 4). Near the bucket–soil contact surface, cell meshes are densely divided, whereas the mesh is sparse in the soil away from the contact surface [20]. Thus, the calculation time is much shortened and the accuracy of the calculation result is improved.

2.1.3. Interaction between Contact Surfaces

To simulate the shear transfer and relative displacement between the suction bucket and the soil, the contact pairs are defined by the bucket–soil surfaces, including inner, outer, top, and bottom surfaces. The master–slave surfaces contact algorithm is used to choose the master surface. The master surfaces are the surfaces of a suction bucket with high stiffness, and the slave surfaces are the surfaces of soil with low stiffness. The contact pair use face-to-face contact and finite sliding. The tangent behavior of the contact pairs is in the form of a Mohr–Coulomb friction penalty function, and the sliding friction coefficient of the interface is $\mu = \tan(0.75\varphi)$ [21] (φ is the internal friction angle of the soil). The normal behavior of the contact pairs is in the form of hard contact, and the suction bucket and soil can be separated.

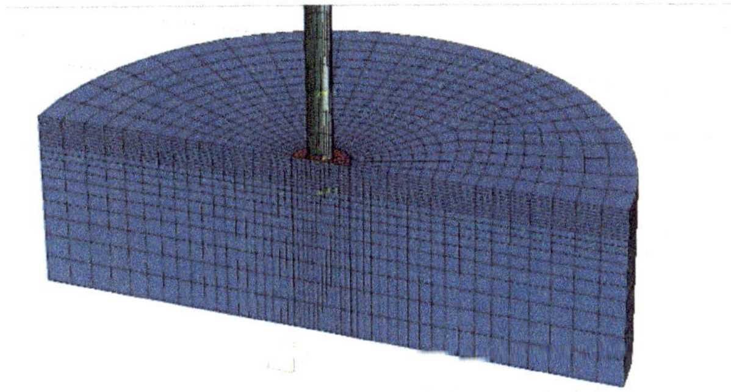

Figure 4. Finite element model.

2.1.4. Initial Geo–Stress Balance

The initial geo–stress was balanced before loading the horizontal cyclic loads by the ODB import method, and the initial geo–stress field was applied to the ground [22]. When the displacement in the direction of Z (U3) reaches 10^{-5}, the additional deformation caused by soil self-gravity is essentially eliminated [23,24].

2.2. Expression of Load

The W3600M-116 offshore wind turbine from China (with 3.6 MW installed capacity) is taken as an example in this paper. H (H = 80 m) is the distance between the hub center and the top surface of the foundation, and G_0 (G_0 = 5200 KN) is the weight of the superstructure transmitted to the foundation. Wind and wave loads are equivalent to horizontal cyclic loads, and cyclic loads are loaded in their respective directions.

Horizontal cyclic loads are defined by periodic amplitude curves in ABAQUS [25]. Fourier series are used to represent periodic amplitude curves.

When $t > t_0$, the amplitude expression is:

$$a = A_0 + \sum_{n=1}^{N} [A_n \cos n\omega(t - t_0) + B_n \sin n\omega(t - t_0)] \quad (1)$$

When $t < t_0$, the amplitude expression is:

$$a = A_0 \quad (2)$$

In Equations (1) and (2), ω is the circular frequency and its value is $\omega = 2\pi f$; f is the frequency; t_0 is the start time; A_0 is the initial amplitude; and A_n is the coefficient of the Cos coefficient and B_n is the Sin coefficient (n = 1, 2, 3 … N).

2.3. Determination of Horizontal Cyclic Load

The most dangerous condition, according to most studies, is when the wind and wave loads on the suction bucket foundations are in the same direction. Wind and wave loads are treated as horizontal cyclic loads exerted on the model in this paper (Figure 5). The magnitude, direction, position, and method of loads exerted are all in accordance with the Load Code for the Design of Building Structures (GB50009-2001) [26] and the data of wind and wave loads for an offshore test site in the East China sea. The height of wind load is 80 m over the seabed surface, the frequency is 0.15 Hz, while the height of wave load is 10 m over the seabed surface, the frequency is 0.1 Hz, according to the specifications and practical engineering. The amplitude of wind load is $F_{\text{wind}} = 334000 \sin(0.3\pi t)$, and the amplitude of wave load is $F_{\text{wave}} = 811000 \sin(0.2\pi t)$.

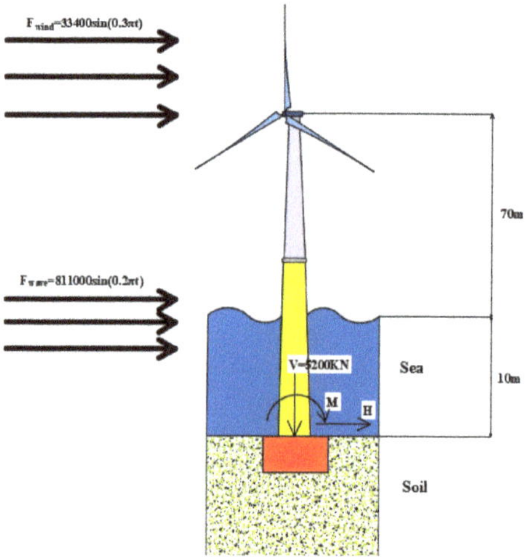

Figure 5. Forces of suction bucket foundation and offshore fan.

2.4. Verification of Finite Element Model

The Equation (3) was used to calculate the relationship between the bearing capacity coefficient and length–diameter ratio (L/D):

$$N_{H,s} = H_{ult,s}/AS_{um} \tag{3}$$

where $N_{H,s}$ is the horizontal ultimate bearing capacity coefficient; $H_{ult,s}$ is the horizontal ultimate bearing capacity; A is sectional area of the suction bucket; and S_{um} is the strength (kPa) of the seabed.

To verify the trustworthiness of the established finite element model, it was compared to existing results [27]. Gelagoti F. et al. set parameters for homogeneous clay, such as L/D = 0.2~1, S_u = 60 kPa, and applied vertical load V = 0 and horizontal force with the method of controlling displacement. In this paper, the same parameters are used, with the exception that the length–diameter ratio (L/D) is increased to 2. When L/D is equal to 1, the calculated results are very close to those found in the literature, with a margin of error of less than 5% (Figure 6). In Figure 6, the horizontal coordinates represent the length-diameter ratio of a suction bucket, and the vertical coordinates represent the horizontal ultimate bearing capacity coefficient.

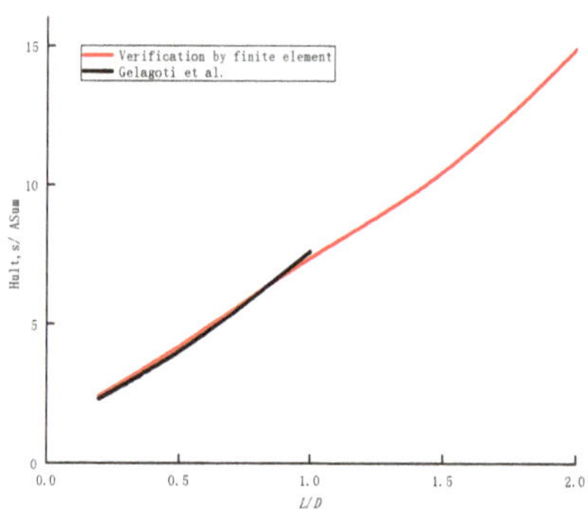

Figure 6. Verification of horizontal bearing capacity coefficients [27].

3. Results

3.1. Bearing Capacity Characteristics with Different Length-Diameter Ratios

The normalized bearing capacity is independent of the diameter of the suction bucket foundation, according to trial calculations and previous studies [28]: suction bucket diameter D = 9 m; suction bucket length L = 4.5, 9, 13.5, and 18 m; length–diameter ratio L/D = 0.5, 1, 1.5, and 2; and suction bucket wall thickness t = 0.1 m.

Assuming that the soil is homogeneous S_{u0} = 5 kPa, the Equation (3) was used to calculate the relationship between the bearing capacity coefficient and length–diameter ratio (L/D). The Figure 7 presents horizontal bearing capacity coefficient with different length–diameter ratios. In Figure 7, the horizontal coordinates represent the length–diameter ratio of a suction bucket, and the vertical coordinates represent the horizontal ultimate bearing capacity coefficient.

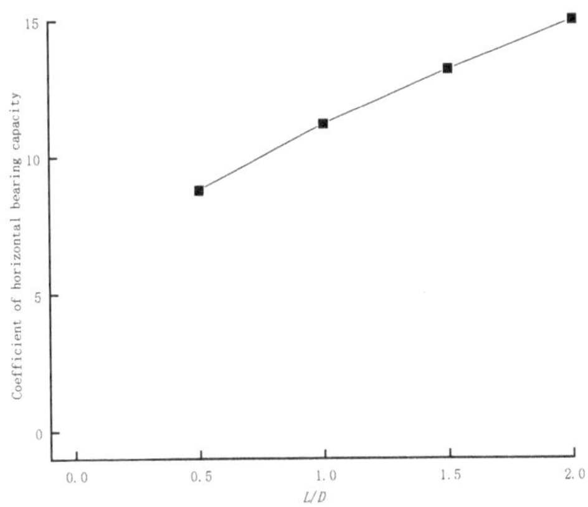

Figure 7. Horizontal bearing capacity coefficient with different L/D.

3.2. Bearing Capacity Characteristics with Different Soil Non-Uniformity Coefficients

Normally consolidated soft clays are widely distributed in the Gulf of Mexico, the South China Sea, and Southeast Asia, and their undrained shear strength S_u increases linearly with depth [29,30], according to Equation (4):

$$S_{um} = KZ + S_{u0} \tag{4}$$

where K is the soil non-uniformity coefficient; Z is the soil depth (m); and S_{u0} is the undrained shear strength of the surface layer. The dimensionless soil non-uniformity coefficient K is used to represent the strength non-uniformity of the soil, according to Equation (5):

$$K = pD/S_{u0} \tag{5}$$

where p is the growth gradient of undrained shear strength along the Z-axis, which is typically 1 to 2.5 kPa/m, p = 1, 1.5, 2, and 2.5 kPa/m are used in this paper; D is the suction bucket diameter (m); and S_{u0} is the undrained shear strength of surface layer.

Five values (K = 0, 1, 2, 5, and 10) are used to investigate the impact of uneven coefficient distribution on bearing capacity characteristics, where K = 0 indicates that the foundation soil is homogeneous soil. The values of K, p, and S_{u0} are shown in Table 3. The horizontal bearing capacity coefficient with different non-uniformity coefficients is presented in Figure 8. In Figure 8, the horizontal coordinates represent the length–diameter ratio of a suction bucket, the vertical coordinates represent the horizontal ultimate bearing capacity coefficient, and the different colored curves represent different soil non-uniformity coefficients.

Table 3. Parameters of K, p, and S_{u0}.

$K = pD/S_{u0}$	p/(kPa·m^{-1})	S_{u0}/kPa
0	0	5
1	1	5
2	1.5	7.5
5	2	4
10	2.5	2.5

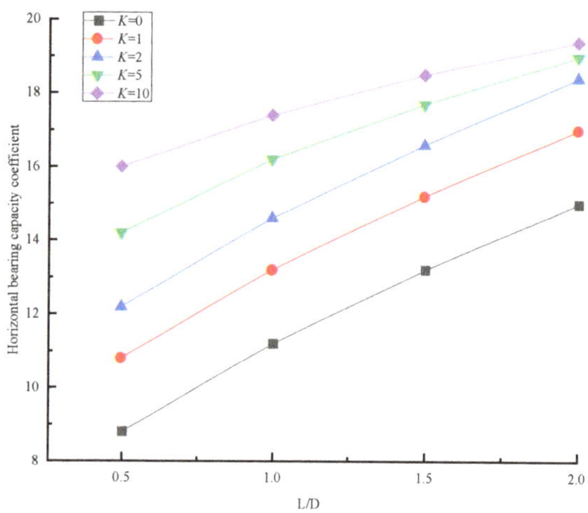

Figure 8. Horizontal bearing capacity coefficient with different non-uniformity coefficients.

3.3. Bearing Capacity Characteristics with Different Angles between Wind and Wave Loads

According to Byrne, B.W., and Houlsby, G.T. [31], the wind and wave load directions are not the same and there is a certain angle between them. According to Seidel [32], the massive horizontal cyclic load induced by wind, wave, and flow does not have a fixed direction, and the included angle of wave load during the service of the offshore fan may exceed 90°. As a result, the wind and wave loads exerted on the offshore fan are not in a single direction, and there is a specific angle θ between them (Figure 9).

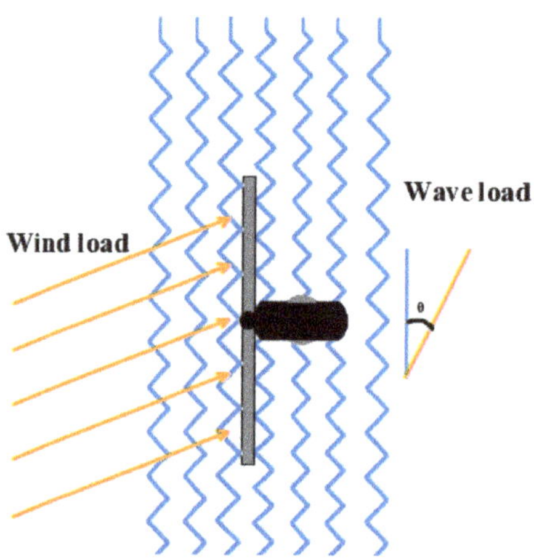

Figure 9. Angles between wind and wave loads.

Generally, the wave load is relatively stable and does not change the direction of action rapidly, whereas turbulent wind causes the wind load to shift the direction rapidly [33]. One-way instability does not occur when the wind and wave loading angle exceeds 45 degrees. As a result, four working conditions of different angles were selected in this paper to investigate the bearing capacity characteristics of the suction bucket. Each working condition is shown in Table 4.

Table 4. Working conditions of wind and wave loading directions.

Working Conditions	Wind Loading Direction	Wave Loading Direction
A	0°	0°
B	0°	15°
C	0°	30°
D	0°	45°

According to the static ultimate bearing capacity and the horizontal displacement corresponding to the ultimate state obtained above, the foundation is judged to have failed when the cyclic displacement amplitude of either of them reaches the ultimate displacement value under the corresponding working condition or if the overall rotation angle of the foundation is 0.05 rad and the corresponding number of cycles when the foundation fails is obtained. The difference in the bearing capacity characteristics of the foundation at different load angles was judged using the number of cycles to failure as an index to compare the bearing characteristics of the foundation. The number of cycles to failure with different wind and wave loading directions is presented in Figure 10. In Figure 10, the horizontal coordinates represent the angle between wind and wave loads, and the vertical coordinates represent the number of cycles in case of foundation failure.

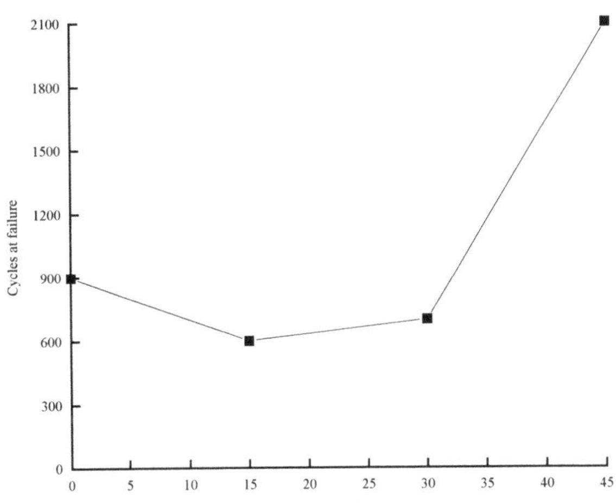

Figure 10. The number of cycles to failure with different wind and wave loading directions.

3.4. *Bearing Capacity Characteristics with Different Number of Cycles*

Considering the fatigue limit state of the offshore fan foundation and the fact that it will undergo approximately 10 million cyclic loads over the course of a 25-year service life [34], the number of cyclic loads will have a significant impact on the accumulated deformation of the foundation, causing the superstructure to reach the normal limit state ahead of schedule [35]. As a result, it is critical to study the impact of the number of cycles on the foundation. Figure 11 presents the attenuation of horizontal bearing capacity after a different number of cycles. In Figure 11, the horizontal coordinates represent the number of cycles and the vertical coordinates represent the horizontal bearing capacity. Figure 12 presents the soil strength weakening at different depths. In Figure 12, the horizontal coordinates represent the percentage of the attenuation of bearing capacity after different number of cycles to the bearing capacity in the initial state and the vertical coordinates represent the depth of the soil.

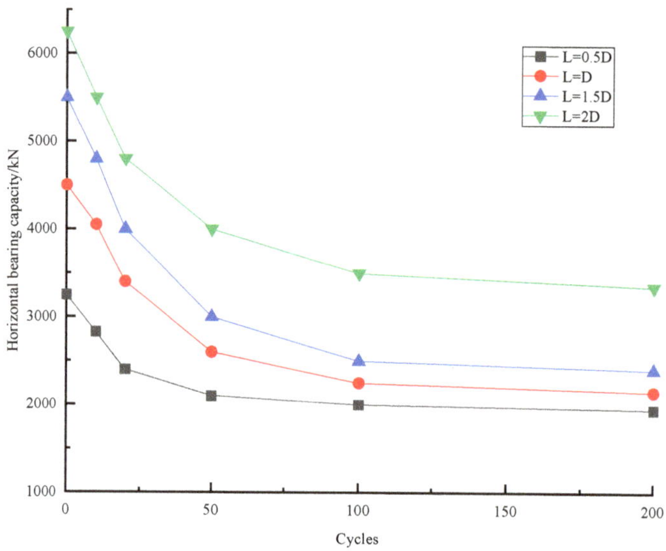

Figure 11. Attenuation of horizontal bearing capacity after different number of cycles.

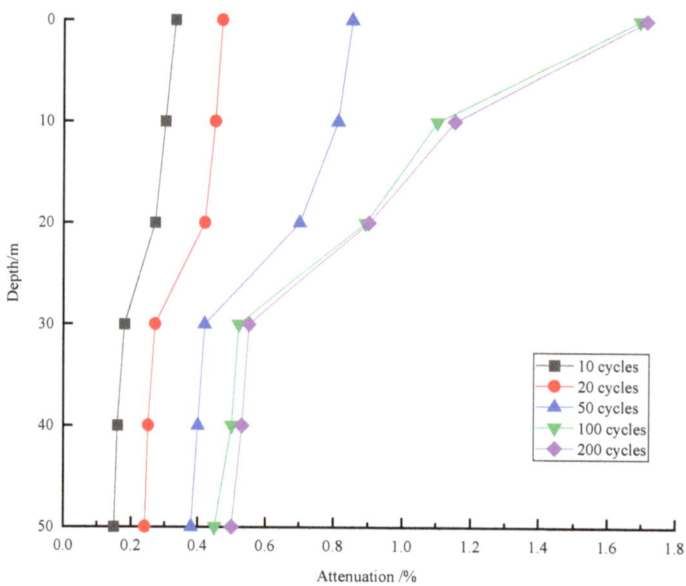

Figure 12. Soil strength weakening at different depths.

4. Discussion

4.1. Sensitivity Analysis of Factor in Length–Diameter Ratios

From the above comparisons of the results from the bearing capacity characteristics with different length–diameter ratios, the horizontal bearing capacity coefficients are respectively calculated while L/D = 0.5, 1.0, 1.5, and 2.0. The horizontal bearing capacity coefficient is 8.8 when L/D = 0.5. The horizontal bearing capacity coefficient increases to 11.2 when L/D is increased to 1.0. The horizontal bearing capacity coefficient increases to 13.2 when L/D is increased to 1.5. The horizontal bearing capacity coefficient increases to 15 when L/D is increased to 2.0. The horizontal bearing capacity coefficient grows constantly as L/D increases, while the rate of increase of bearing capacity drops progressively from 4.8 to 4 and 3.6. When the length–diameter ratio is small, the horizontal bearing capacity coefficient increases with the increase of the length–diameter ratio, but the rate of increase slows down. The bearing capacity coefficient of the suction bucket remains essentially unaltered when the length–diameter ratio is reasonably big and reaches a specific depth. In other words, the buried depth only affects the suction bucket's bearing capacity characteristics within a particular range. The impact is minimal when the threshold is exceeded.

4.2. Sensitivity Analysis of Factor in Soil Non-Uniformity Coefficients

From the above comparisons of the results from the bearing capacity characteristics with different soil non-uniformity coefficients, the horizontal bearing capacity coefficients (K = 0, 1, 2, 5, and 10) are respectively calculated. When the length–diameter ratio is 0.5, the horizontal bearing capacity coefficients (K = 0, 1, 2, 5, and 10) are 8.8, 10.8, 12.2, 14.2, and 16 respectively. It indicates that the horizontal bearing capacity coefficient increases as the non-uniformity coefficient K increases. When the length–diameter ratio is 2.0, the horizontal bearing capacity coefficients (K = 0, 1, 2, 5, 10) are 17, 18.4, 19, 19.4, respectively. It verifies that the influence of soil unevenness coefficient on the bearing capacity coefficients is greater than the influence of the length–diameter ratio when the length–diameter ratio is small, while the length–diameter ratio is large enough, the influence of soil non-uniformity coefficient on the bearing capacity of the suction bucket foundation gradually diminishes.

4.3. Sensitivity Analysis of Factor in Angles between Wind and Wave Loads

From the above comparisons of the results from the bearing capacity characteristics with different angles between wind and wave loads, when the wind and wave loading angle is 15° to 30°, the bearing capacity characteristic of the foundation is weakened to some extent compared to the loading direction collinearity. The reason for this is that the type of load exerted on the foundation varies. Although the value is minor, the wind load causes a substantial bending moment in the foundation, making the bearing characteristics in the direction of about 15° to 30° weaker than in the direction of 0°. Furthermore, the lateral displacement of the wave load promotes the foundation's instability. The results show that a particular load angle causes the foundation's bearing capacity to deteriorate, putting the suction bucket foundation in jeopardy.

4.4. Sensitivity Analysis of Factor in Number of Cycles

As can be seen from the above comparisons of the results and from the bearing capacity characteristics with different number of cycles, the horizontal bearing capacity of a suction bucket foundation initially decreases rapidly as the number of cycles increases. The horizontal bearing capacity had decreased by more than 95% after 100 cycles. The weakening of soil strength is related to the attenuation of horizontal bearing capacity after cyclic loading. The plastic strain will appear after the soil has been subjected to cyclic load. The accumulation of plastic strain leads to soil strength weakening, and the degree of soil strength weakening decreases gradually as plastic strain accumulates.

The soil stress at different depths decreases with the increase in the number of cycles after the suction bucket is subjected to the horizontal cyclic load, and the shallow soil stress decreases much more than deep soil. After about 100 cycles, the soil stress at shallow gradually increases and is basically stable, whereas the soil stress at deep continues to decrease but the amplitude becomes low. As a result, when a suction bucket is subjected to cyclic loading, the strength of shallow soil deteriorates rapidly because it serves as the primary source of horizontal bearing capacity, whereas the strength of deep soil deteriorates slowly. The strength of deep soil causes attenuation even when the horizontal bearing capacity is essentially stable.

5. Conclusions

A numerical model of the interaction between the suction bucket and saturated soft clay was established. The sensitivity of different factors affecting the bearing capacity of the suction bucket foundation, including length–diameter ratios, soil heterogeneity, angles between wind and wave loads, and the number of cycles was analyzed. Based on the above results and discussion, the following conclusions can be drawn:

1. The relationship between the horizontal bearing capacity coefficient and the length-diameter ratio and the soil unevenness coefficient has been established through numerical simulation, and the results show that the length–diameter ratio and the unevenness coefficient are positively correlated to the horizontal bearing capacity coefficient.
2. When the length–diameter ratio of suction bucket is less than 1.0, the influence of the soil non-uniformity coefficient on the bearing capacity of the suction bucket foundation is greater than length–diameter ratio.
3. When the length–diameter ratio of the suction bucket is more than 2.0, the influence of soil non-uniformity coefficient on the bearing capacity of the suction bucket foundation gradually diminishes.
4. When the angle between wind and wave loads is 15° to 30°, the bearing capacity of the suction bucket foundation is lower than the bearing capacity when the wind and wave loads are collinear.
5. The attenuation of soil strength is rapid within 60 cycles. Because the shallow soil layer is the primary provider of horizontal bearing capacity during the initial cycles, the rate of strength attenuation in the shallow soil is faster than that in the deep soil under initial cyclic loads.
6. The attenuation of soil strength is slow after 100 cycles. Because the strength of shallow soil is relatively stable after 100 cycles, the reason for the subsequent slow strength attenuation is that the strength of deep soil decreases.

Author Contributions: Conceptualization, B.W. and C.-F.Y.; methodology, L.L.; software, M.-H.Y.; validation, M.-H.Y. and C.-F.Y.; formal analysis, L.L.; data curation, Y.L.; writing—original draft preparation, B.W. and M.-H.Y.; writing—review and editing, C.-F.Y. and L.L.; visualization, K.-M.S.; funding acquisition, B.W. All authors have read and agreed to the published version of the manuscript.

Funding: This research was funded by the National Natural Science Foundation of China (Grant No. 51909238, 51939002, 52101334, 52071301), the Open Fund of Zhejiang Provincial Key Laboratory of Deep-sea Wind Power Technology Research (Grant No. ZOE2020006) and Zhejiang Provincial Natural Science Foundation of China (LHY21E090001).

Institutional Review Board Statement: Not applicable.

Informed Consent Statement: Not applicable.

Data Availability Statement: Not applicable.

Acknowledgments: The authors would like to acknowledge the funding for support. Finally, we would also like to thank the associate editor of this Special Edition, and the reviewers who provided constructive feedback that has without doubt improved this manuscript.

Conflicts of Interest: Authors B.W. and K.-M.S. are employed by Powerchina Huadong Engineering Corporation Limited. The remaining authors declare that the research was conducted in the absence of any commercial or financial relationships that could be construed as a potential conflict of interest.

References

1. Zou, L.; Tang, H.Y.; Wang, Y.H.; Chen, P.; Huang, X.G.; Wu, W. Development of Foundation Structure Selection Program of Offshore Wind Turbines. *Ship Eng.* **2020**, *42*, 282–285. (In Chinese)
2. Sun, Y.X. Experimental and Numerical Studies on a Laterally Loaded Monopile Foundation of Offshore Wind Turbine. Ph.D. Thesis, Zhejiang University, Zhejiang, China, 2016. (In Chinese).
3. Xie, M.; Lopez-Querol, S. Numerical Simulations of the Monotonic and Cyclic Behavior of Offshore Wind Turbine Monopile Foundations in Clayey Soils. *J. Mar. Sci. Eng.* **2021**, *9*, 1036. [CrossRef]
4. Amin, B.; Britta, B.; Domenico, L.; Shinji, S. Offshore Wind Turbine Foundations. *Soils Found.* **2021**, *61*, 621–622.
5. Li, J.W.; Hu, G.Q.; Jin, G.Q. Hydrodynamic Performance of a Novel Floating Foundation for an Offshore Wind Turbine Under a Storm Condition. *Int. J. Offshore Polar Eng.* **2020**, *30*, 120–128. [CrossRef]
6. Tjelta, T.L. Geotechnical Aspects of Bucket Foundations Replacing Piles for the Europipe 16/11-E Jacket. In Proceedings of the Offshore Technology Conference, Houston, TX, USA, 2 May 1994.
7. Shi, X.C.; Xu, R.Q.; Gong, X.N.; Chen, G.X.; Yuan, Z.L. Introduction of Bucket Foundation. *Chin. Civil Eng. J.* **2000**, *33*, 68–92. (In Chinese)
8. Yang, J.L.; Wang, S.F.; Kong, L.W.; Yuan, J.X.; Liu, Z.A.; Wang, Q. The 3-D Finite Element Analysis of Stability of Platform with Bucket Foundation. *Rock Soil Mech.* **2002**, *23*, 640–644. (In Chinese)
9. Wu, K.; Xue, H.F.; Chen, R.; Li, S.C. Studies on Bearing Capacity of Composite Multi-Suction Bucket Structure of Suctional Suction Bucket Foundation. *J. Disaster Prev. Mitigation Eng.* **2008**, *28*, 484–491. (In Chinese)
10. Li, D.Y.; Chen, Q.J.; Zhang, Y.K.; Chen, F.Q. Review of Horizontal Loading on Suction Caisson of Offshore Wind Turbines. *Ocean Eng.* **2020**, *38*, 137–147. (In Chinese)
11. Wu, X.N.; Liao, Q.; Li, Y. A Review on the Bearing Capacity Studies of the Suction Bucket Foundation for Offshore Wind Turbines. *J. Ocean Technol.* **2020**, *39*, 91–106. (In Chinese)
12. Jin, S.C.; Zhang, Y.T.; Yang, Y.H.; Li, B. Research on Horizontal Ultimate Bearing Capacity of Suction Bucket Foundation in Saturated Sand Ground in Saturated Sand Ground. *Rock Soil Mech.* **2013**, *34*, 221–227. (In Chinese)
13. Sun, X.Y.; Luan, M.T.; Tang, X.W. Study of Horizontal Bearing Capacity of Bucket Foundation on Saturated Soft Clay Ground. *Rock Soil Mech.* **2010**, *31*, 667–672. (In Chinese)
14. Wu, K.; Luan, M.T.; Fan, Q.L.; Wang, Z.Y. Numerical Analysis of Failure Envelopes of Bucket Foundation Subjected to Combined Loads Based on Elasto-plastic Fem. *Eng. Mech.* **2008**, *25*, 156–161. (In Chinese)
15. Zhang, Y.K.; Wang, C.C.; Li, D.Y.; Qi, Y.S. Numerical Simulation of Bearing Characteristics of Modified Suction Cassions in Sand Under Combined Loads. *Sci. Technol. Eng.* **2021**, *21*, 1515–1521. (In Chinese)
16. Wang, J.H.; Yang, H.M. Model test on Horizontal Cyclic Bearing Capacity of Bucket Foundations in Soft Clays. *Rock Soil Mech.* **2008**, *29*, 2606–2612. (In Chinese)
17. Zhao, S.X.; Sun, B.B. *Code for Design of Wind Turbine Foundations for Offshore Wind Power Projects (NB/T 10105-2018)*, 1st ed.; Yi, Y.C., Xie, H.W., Eds.; China Water & Power Press: Beijing, China, 2019; pp. 3–4.
18. Hong, Z.B.; Tao, J.G.; Hu, D.; Li, F.; Xie, X.T. Research on the Stiffness Degradation Model of Saturated Soft Clay Seabed under Cyclic Loading. *J. Railw. Sci. Eng.* **2021**, *18*, 351–358. (In Chinese)
19. Fan, Q.L. A Study on Stability of Deeply-Embedded Large-Diameter Cylindrical Structure in Soft Ground. Ph.D. Thesis, Dalian University of Technology, Liaoning, China, 2007. (In Chinese).
20. Chen, G.; Beer, M.; Liu, Y. Modeling Response Spectrum Compatible Pulse-like Ground Motion. *Mech. Syst. Signal Proc.* **2022**, *177*, 109177. [CrossRef]
21. Liu, J.C.; Xiong, G.; Zhu, B.; Ying, P.P. Bearing Capacity and Deflection Behaviors of Large Diameter Monopile Foundations in Sand Seabed. *Rock Soil Mech.* **2015**, *36*, 591–599. (In Chinese)
22. Liu, Y.; Jiang, Y.; Xiao, H.; Lee, F.H. Determination of Representative Strength of Deep Cement-Mixed Clay from Core Strength Data. *Géotechnique* **2017**, *67*, 350–364. [CrossRef]
23. Liu, Y.; Lee, F.H.; Quek, S.T.; Chen, E.J.; Yi, J.T. Effect of Spatial Variation of Strength and Modulus on the Lateral Compression Response of Cement-Admixed Clay Slab. *Géotechnique* **2015**, *65*, 851–865. [CrossRef]
24. Wang, R.H.; Li, D.Q.; Chen, E.J.; Liu, Y. Dynamic Prediction of Mechanized Shield Tunneling Performance. *Autom. Constr.* **2021**, *132*, 103958. [CrossRef]
25. Cao, J.F.; Shi, Y.P. *Answers to Common Issues of Finite Element Analysis in ABAQUS*, 1st ed.; Kong, J., Guo, J., Eds.; China Machine Press: Beijing, China, 2009; pp. 109–114.
26. Jin, X.Y.; Wang, J. *Load Code for the Design of Building Structures (GB50009-2012)*, 1st ed.; Chen, M.K., Wang, D.S., Eds.; China Architecture Publishing & Media Co., Ltd.: Beijing, China, 2012; pp. 30–62.
27. Gelagoti, F.; Georgiou, I.; Kourkoulis, R.; Gazetas, G. Nonlinear Lateral Stiffness and Bearing Capacity of Suction Caissons for Offshore Wind-Turbines. *Ocean Eng.* **2018**, *170*, 445–465. [CrossRef]
28. Zhou, S.J.; Zhang, Y.; Wang, D. Capacity Envelope of Caisson Foundation for on-bottom Pipelines. *Mar. Sci. Bull.* **2019**, *38*, 727–732. (In Chinese)
29. Chen, X.J.; Li, D.Q.; Tang, X.S.; Liu, Y. A Three-Dimensional Large-Deformation Random Finite-Element Study of Landslide Runout Considering Spatially Varying Soil. *Landslides* **2021**, *18*, 3149–3162. [CrossRef]

30. Liu, Y.; He, L.Q.; Jiang, Y.J.; Sun, M.M.; Chen, E.J.; Lee, F.H. Effect of in-Situ Water Content Variation on the Spatial Variation of Strength of Deep Cement-Mixed Clay. *Géotechnique* **2019**, *69*, 391–405. [CrossRef]
31. Byrne, B.W.; Houlsby, G.T. Foundations for Offshore Wind Turbines. *Philos. Trans. R. Soc. Lond. Ser. A* **2003**, *361*, 2909–2930. [CrossRef]
32. Seidel, M. Feasibility of Monopiles for Large Offshore Wind Turbines. In Proceedings of the 10th German Wind Energy Conference (DEWEK), Bremen, Germany, 17–18 November 2010.
33. Fischer, T.; Rainey, P.; Bossanyi, E.; Kühn, M. Study on Control Concepts Suitable for Mitigation of Loads from Misaligned Wind and Waves on Offshore Wind Turbines Supported on Monopiles. *Wind Eng.* **2011**, *35*, 561–574. [CrossRef]
34. Schaumann, P.; Lochte-Holtgreven, S.; Steppeler, S. Special Fatigue Aspects of Support Structures with Offshore Wind Turbines. *Materialwiss. Werkstofftech.* **2011**, *42*, 1075–1081. [CrossRef]
35. Liu, Y.; Li, K.Q.; Li, D.Q.; Tang, X.S.; Gu, S.X. Coupled Thermal–Hydraulic Modeling of Artificial Ground Freezing with Uncertainties in Pipe Inclination and Thermal Conductivity. *Acta Geotech.* **2022**, *17*, 257–274. [CrossRef]

sustainability

MDPI

Article

Experimental Study on the Preparation of a Highly Active Bacterial Suspension for MICP in the South China Sea

Yuxuan Zhou [1], Jun Hu [1], Yuwei Wu [1,*], Hui Zeng [2], Dongling Zeng [3] and Zhixin Wang [3]

1 School of Civil Engineering and Architecture, Hainan University, Haikou 570228, China
2 School of Civil Engineering and Architecture, Wuyi University, Jiangmen 529020, China
3 Hainan Investigation Institute of Hydrogeology and Engineering Geology, Haikou 570206, China
* Correspondence: wuyuwei@eis.hokudai.ac.jp

Abstract: Most previous studies investigating the use of microbial-induced calcium carbonate precipitation (MICP) to reinforce foundations have indicated that the final curing effect can be improved by changing the nutrient environment parameters during the MICP reaction. However, using such methods to improve the construction process increases the construction cost and the impact on the surrounding environment. This study was conducted to determine if we could prepare a bacterial suspension with high activity in a short period of time by examining the effects of inoculation time, the concentration of the inoculated bacterial suspension, and shaker speed during expansion to determine whether sporosarcina pasteurii would vigorously grow. Based on the analysis of the pattern of activity variations in the bacterial suspension, the most appropriate growth scheme for preparing high-activity bacterial suspensions when using MICP to strengthen structures in the South China Sea was proposed. In terms of the results, it was found that the trend of changes in activity and the peak time of maximum activity in the bacteriophage cultured under low-speed conditions in the expanded culture tended to be the same. The value of the bacteriophage's activity was low. During medium speed culture of the bacterial suspension, urease activity peaked much higher than that of the other comparison groups, with the medium speed bacterial broth having the highest peak. As a result of the prolonged shaking incubation time, the effect of prolonged shaking on urease activity in the bacterial suspension was mainly reflected in the fact that the activity decay cycle of the colony itself was slowed.

Keywords: South China Sea environment; microbial-induced carbonate precipitation; training program; urease activity; change rule

1. Introduction

Due to the policy of "One Belt, One Road," China is paying increasingly more attention to developing and constructing islands and reefs in the South China Sea. Calcareous sand is widely distributed in the South China Sea, and its particles are irregular in shape, rich in micropores, and easily broken, which easily causes uneven settlement of the foundation; therefore, it should be appropriately treated during its development and construction [1,2]. Despite the maturity of traditional foundation treatment methods (such as vibratory punching), they do not apply to calcareous sand foundations, and the construction is difficult, expensive, and not environmentally friendly.

In recent years, MICP technology has emerged as a research hotspot in civil engineering at home and abroad, and is widely used for sand reinforcement [3–6], restoration of ancient sites [7–10], and remediation of contaminated soils [11–14]. The advantages of this type of construction are its ease of construction, low cost, and environmental friendliness. The main mechanism of reinforcement is as follows: in its metabolism, sporosarcina pasteurii produces urease, which hydrolyzes urea in the environment to form carbonate, which combines with calcium ions in the environment to create $CaCO_3$ crystals that can function as cement. The main chemical reaction equations can be summarized as follows [15–17].

$$CO(NH_2)_2 + 2H_2O \rightarrow H_2CO_3 + 2NH_3$$

$$H_2CO_3 + 2NH_3 \leftrightarrow 2NH_4^+ + 2OH^-$$

$$H_2CO_3 \rightarrow H^+ + HCO_3^-$$

$$HCO_3^- + 2OH^- + H^+ \leftrightarrow CO_3^{2-} + 2H_2O$$

$$O_3^{2-} + Ca^{2+} \leftrightarrow CaCO_3 \downarrow$$

Researchers Masahiro Fujita et al. [18,19] found that the high pH of the solution affected the growth of bacteria, resulting in a weaker curing effect of MICP. As determined by Lv et al. [20], calcium carbonate precipitates were strongly influenced by temperature, solution pH, bacterial suspension concentration, and colloidal solution concentration, with the size of the resulting calcium carbonate crystals increasing with temperature. Liang et al. [21] investigated the effect of cement content on the physical and mechanical properties of reinforced ISO standard sand specimens. Tan et al. [22] demonstrated that binder concentrations less than 1 mol/L had a good effect on geotechnical properties. At a cement content of 0.25 mol/L, the permeability and/or long-term stability of geotechnical soil was significantly increased, however, it took longer, while at a cement content of 0.5 mol/L, the calcium carbonate content and unconfined compressive strength of the specimens were relatively high. Al Thawadi [23] and Lai et al. [24] found that high cement concentrations delayed or even terminated the MICP process, with the delay effect gradually accounting for the obvious when cement concentrations exceeded 1 mol/L. Cement concentrations up to 2.5 mol/L will lead to the termination of the calcium carbonate production process during MICP. Han et al. [25] investigated the effect of three different calcium salt–urea nutrient salt solutions on the consolidation of liquefiable sand by MICP and found that a $Ca(CH_3COO)_2$-urea nutrient salt consolidation was the most effective. Zhao et al. [26] found that $Ca(CH_3COO)_2$ was more suitable as a calcium source than $CaCl_2$ or $Ca(NO_3)_2$, and its pore size distribution was wider after consolidation by the MICP process. The specimens had a wider range of pore sizes and higher strength. Cheng et al. [27] investigated the effect of four calcium sources and Ca^{2+} concentrations on the induction of calcium carbonate precipitation by autonomously screened UPB1 bacteria, with $CaCl_2$ as the calcium source producing the most stable rhomboid-type calcium carbonate with the highest deposition efficiency at a Ca^{2+} concentration of 0.25 mol/L. In a study by Maleki Kakelar et al. [28], using different injection steps (reaction time, fixative, and number of injections), the sand sample strength increased from 0.25 MPa to 1.55 MPa when the calcium carbonate content increased from 7.7% to 18.9%. Typically, MICP studies are carried out under nutrient-sufficient conditions. Yusuf Atay Erş an et al. [29] used a special elastomeric strain to increase the yield of $CaCO_3$ precipitation in low-nutrient environments, which greatly improved the efficiency of nutrient utilization in the environment.

Even though most of the existing studies alter the nutrient environment parameters of the MICP construction process in order to increase its performance, such alterations may increase construction costs and have adverse effects on the surrounding environment, such as imbalances in the proportion of elements in the environment and eutrophication of the water column. This study investigated the origin of the MICP process (i.e., the ability of bacteria to hydrolyze urea), developed a highly active bacterial suspension in a short time by optimizing the culture protocol of sporosarcina pasteurii, and performed variation pattern analysis to determine the most suitable conditions for preparing an active bacterial suspension for MICP in the South China Sea. The growth, reproduction, and metabolism of microorganisms are closely related to external factors and are influenced by environmental conditions. To reduce the influence of environmental factors on the test results and enhance the project's practicality, the test site was chosen in Haikou, China.

2. Materials and Methods

Before preserving strains may be used for expanded cultures, they must first be activated and stimulate their bacterial activity. Current preservation methods include liquid paraffin, sand preservation, vacuum freeze drying, preservation in liquid nitrogen, and subculturing [30]. The strains used in this study were maintained primarily through subculturing.

The test procedure is illustrated in Figure 1. The liquid medium needs to be prepared first. After sterilization in an autoclave, the prepared medium is cooled on a sterile bench until it is ready for use. As soon as the preserved culture medium has been taken out from the refrigerator, it is brought to room temperature on the sterile bench. The room-temperature medium is inoculated on the aseptic table and the medium sealed to prevent contamination. After inoculation, the broth is incubated in a constant temperature shaker. It is important to note that the activity of the strain is tested after incubation, and the activation process is repeated until the activity of the bacterial suspension is fully stimulated. After activating the strain, the nutrient solution is added to the liquid medium in a set proportion for incubation and the temperature and speed of the shaker are set according to the experimental protocol.

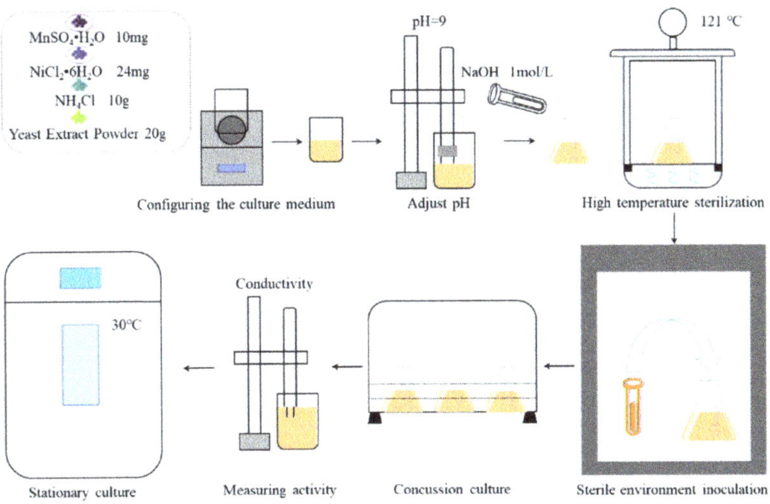

Figure 1. Schematic of the experimental procedure.

The strain used in the test was sporosarcina pasteurii, a Gram-positive bacterium purchased from the General Microbial Strain Repository of China. In geotechnical engineering, this bacterium is the most commonly used and is typically ovoid and approximately 1 mm to 3 mm in size.

Based on the study by Chu et al. [31–33], the parameters of the media used in this experiment are listed in Table 1. The solid and liquid media were used in this experiment. The solid medium is used for the long-term preservation of the strains and the liquid medium is used for the activation and expansion of the bacteria. In this case, the solid medium had the same nutrient composition as the liquid medium, differing only in the inclusion of agar for solidification.

Figure 2 shows the experimental apparatus and the experimental operations used during the experiment.

Wiffin et al. [34] found a linear relationship between the amount of urea hydrolyzed in solution and the value of conductivity change so that conductivity values may be used as an indirect index of urease activity. A bacterial suspension of 2 mL was added to 18 mL of urea solution containing a concentration of 1.5 moles per liter, and the mixture was stirred thoroughly.

Table 1. Composition and proportion of culture medium.

Ingredients	Content/L
$MnSO_4 \cdot H_2O$	10 mg
$NiCl_2 \cdot 6H_2O$	24 mg
NH_4Cl	10 g
Yeast extract	20 g
Agar powder	15 g (for solid media)

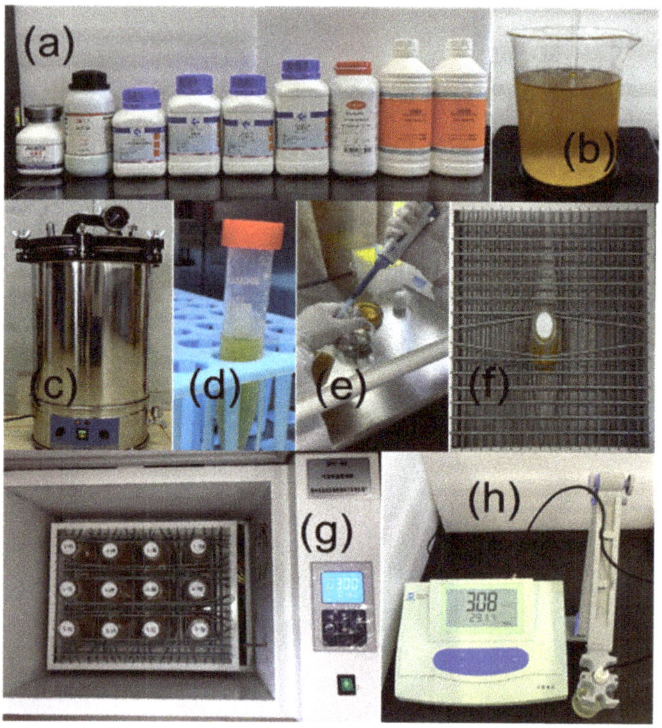

Figure 2. Experimental process diagram. (**a**) Preparation of chemicals to be used; (**b**) configuration of liquid medium; (**c**) sterilization of liquid culture media; (**d**) dispensation of liquid medium and cooling to room temperature; (**e**) inoculation of bacterial suspension; (**f**) stationary culture; (**g**) shaking of culture; (**h**) detection of urease activity.

The conductivity change values were measured over 5 min using a conductivity meter, and the actual conductivity change (mS/cm/min) was used to characterize the urease activity. Where the following equation can express the relationship between the rate of change in conductivity and urease activity,

$$U = C \times 11.11$$

$$U_{AR} = \frac{C_6 - C_1}{5} \times 10 \times 11.11$$

$$U_A = \frac{C_6 - C_1}{5} \times 10$$

where U is the amount of urea hydrolyzed (mM); C is the conductivity change value (mS/cm); U_{AR} is the actual urea hydrolysis rate (mM urea/min); C_6 and C_1 are the conductivity values at minute 6 and minute 1, respectively; and U_A is the urease activity (mS/cm/min).

Three parallel groups were tested using the gradient method, and the mean value of the urease activity was used as the standard value to characterize the results. The colonies were expanded under different experimental conditions, and the level of urease activity was assessed at the end of the experiment. Experimental conditions included the volume of inoculation (1%, 2%, 3%), rotational speed (160 rpm, 190 rpm, 220 rpm), and time of shaking in cultivation (12 h, 24 h, 36 h, 48 h). The effect of bacteria broth prepared for a period of time (12 h, 24 h, 36 h) was studied at 30 °C. The specific experimental protocols are summarized in Table 2. This study aimed to assess the most appropriate culture protocols and recommendations for using highly active bacterial suspensions in the environment of the South China Sea for future applications.

Table 2. Experimental conditions.

Experimental Number	Inoculation Volume (%)	Rotational Speed (rpm)	Cultivation Time (h) 48 h in Total		Temperature (°C)	Interval of Activity Measurement (h)
			Shake Culture (h)	Stationary Culture (h)		
$A_{1\text{-}12}$	1	160	12	36	30	12
$A_{1\text{-}24}$			24	24		
$A_{1\text{-}36}$			36	12		
$A_{1\text{-}48}$			48	0		
$A_{2\text{-}12}$	2		12	36		
$A_{2\text{-}24}$			24	24		
$A_{2\text{-}36}$			36	12		
$A_{2\text{-}48}$			48	0		
$A_{3\text{-}12}$	3		12	36		
$A_{3\text{-}24}$			24	24		
$A_{3\text{-}36}$			36	12		
$A_{3\text{-}48}$			48	0		
$B_{1\text{-}12}$	1	190	12	36	30	12
$B_{1\text{-}24}$			24	24		
$B_{1\text{-}36}$			36	12		
$B_{1\text{-}48}$			48	0		
$B_{2\text{-}12}$	2		12	36		
$B_{2\text{-}24}$			24	24		
$B_{2\text{-}36}$			36	12		
$B_{2\text{-}48}$			48	0		
$B_{3\text{-}12}$	3		12	36		
$B_{3\text{-}24}$			24	24		
$B_{3\text{-}36}$			36	12		
$B_{3\text{-}48}$			48	0		

Table 2. *Cont.*

Experimental Number	Inoculation Volume (%)	Rotational Speed (rpm)	Cultivation Time (h) 48 h in Total		Temperature (°C)	Interval of Activity Measurement (h)
			Shake Culture (h)	Stationary Culture (h)		
$A_{1\text{-}12}$	1	220	12	36	30	12
$A_{1\text{-}24}$			24	24		
$A_{1\text{-}36}$			36	12		
$A_{1\text{-}48}$			48	0		
$A_{2\text{-}12}$	2		12	36		
$A_{2\text{-}24}$			24	24		
$A_{2\text{-}36}$			36	12		
$A_{2\text{-}48}$			48	0		
$A_{3\text{-}12}$	3		12	36		
$A_{3\text{-}24}$			24	24		
$A_{3\text{-}36}$			36	12		
$A_{3\text{-}48}$			48	0		

3. Results

The purpose of this chapter is to discuss the effects of different experimental conditions on the expanded culture of the bacteriophage. The effect of different experimental environments on bacterial culture activity was investigated using single factor analysis. The control experiments in the article were set up with three parallel control groups with the same experimental conditions, and the results of the three experimental groups were plotted in a line graph. The upper end of the error bar represents the maximum value of the measured experimental results in the three groups, and the lower end of the error bar is the minimum value of the experimental results in the three groups, and the average value of the three experimental results is taken as the characteristic value.

3.1. Test Results

3.1.1. Effect of Inoculum Volume on Urease Activity

As shown in Figure 3, the trend of urease activity U_A and the time point of peak activity were both consistent, the effect of inoculum level on peak activity was not significant, and the stability of peak activity was better when a 3% inoculum volume was selected. In both the shaking and static incubations, the peak urease activity was reached after 24 h (up to 0.36 mS/cm/min). In general, the higher the inoculum volume, the faster the decay of urease activity.

As can be seen in Figure 4, the pattern of variation in urease activity between bacterial suspensions varied based on the inoculum used. The urease activity was 2% higher than 1% if the culture was shaken only for 12 h. The urease activity decayed rapidly if shaking was not continued for 12 h. As to this point, the 1% inoculum had slightly less decay, but the activity would remain stable if left to stand. If the shaking was continued for a period of 12 h, the activity of the 1% inoculum tended to decrease, whereas in the other two groups, the opposite is true. With increasing inoculum levels, urease activity changed from increasing to decreasing after 24 h of shaking if the broth was left to stand. When shaking was continued, urease activity increased in all groups, and the higher the inoculum level, the more significant the increase.

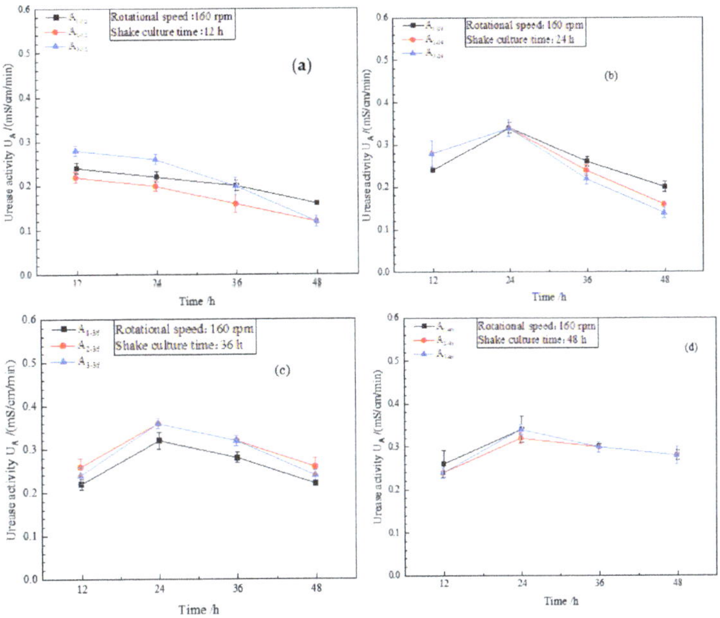

Figure 3. Variation in urease activity with rotational speed of 160 rpm at (**a**) 12 h; (**b**) 24 h; (**c**) 36 h; (**d**) 48 h.

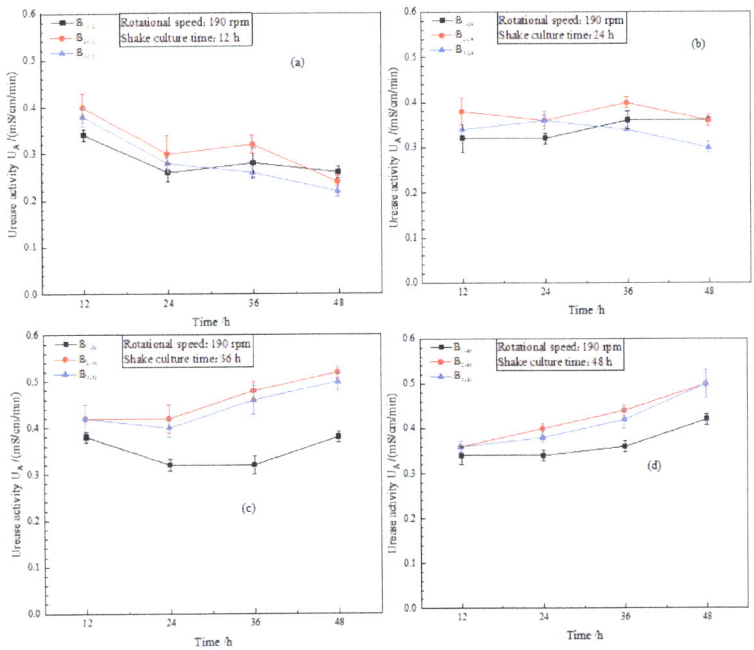

Figure 4. Variation in urease activity with rotational speed of 190 rpm at (**a**) 12 h; (**b**) 24 h; (**c**) 36 h; (**d**) 48 h.

As shown in Figure 5, the activity of different inoculum volumes varies with time during vacuum incubation at a speed of 220 rpm. As can be observed from Figure 5, the inoculum volume did not have much influence on the shaking culture when the speed was 220 rpm. During the shaking culture and the 12 h of resting time for each group, urease activity was always maintained within a certain range and stable. The activity of urease varied with time and inoculum amount, with the highest activity (0.32 mS/cm/min) at 1% inoculum after 36 h.

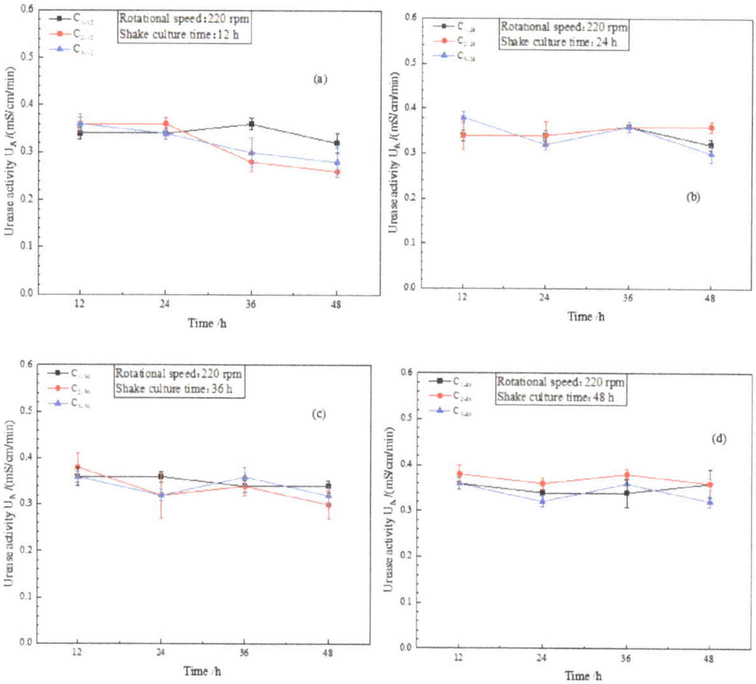

Figure 5. Variation in urease activity with rotational speed of 220 rpm at (**a**) 12 h; (**b**) 24 h; (**c**) 36 h; (**d**) 48 h.

3.1.2. Effect of Rotational Speed on Urease Activity

As shown in Figure 6, the urease activity varies with time when incubated at different rotational speeds. The urease activity increased with the increase in rotation speed in all time periods. The activity of the bacteria cultured at high speed (220 rpm) remained stable after resting, the activity of the bacteria cultured at medium speed (190 rpm) decreased and then stabilized after resting, and the activity of the bacteria cultured at low speed (160 rpm) continued to decay slowly after resting. The urease activity of all groups increased to varying degrees with increasing shaking time after 2 days of incubation. The urease activity of the medium and high-speed cultures was more stable, and the medium-speed culture was slightly better, while the urease activity of the low-speed shaking culture began to decay after reaching its peak at 24 h, and the activity decayed more quickly under rest.

In Figure 7a, the values of urease activity were 190 rpm, 220 rpm, and 160 rpm, respectively, and in the medium and high-speed shaking cultures, values of urease activity were closer. Following 36 h or 48 h of shaking the cultures, the values of urease activity were similar in low and high-speed cultures, but not as good as the medium-speed culture.

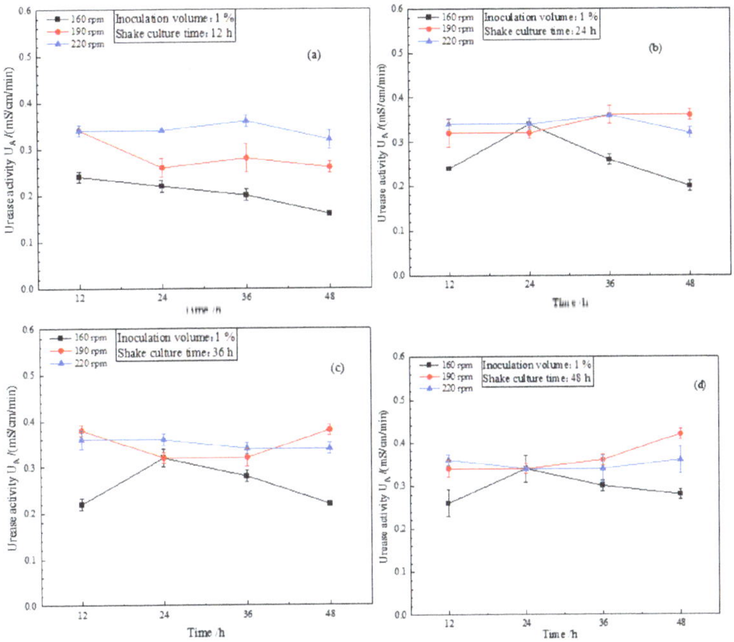

Figure 6. Urease activity after inoculation with 1% of the bacterial suspension with different shaking culture times at (**a**) 12 h; (**b**) 24 h; (**c**) 36 h; (**d**) 48 h.

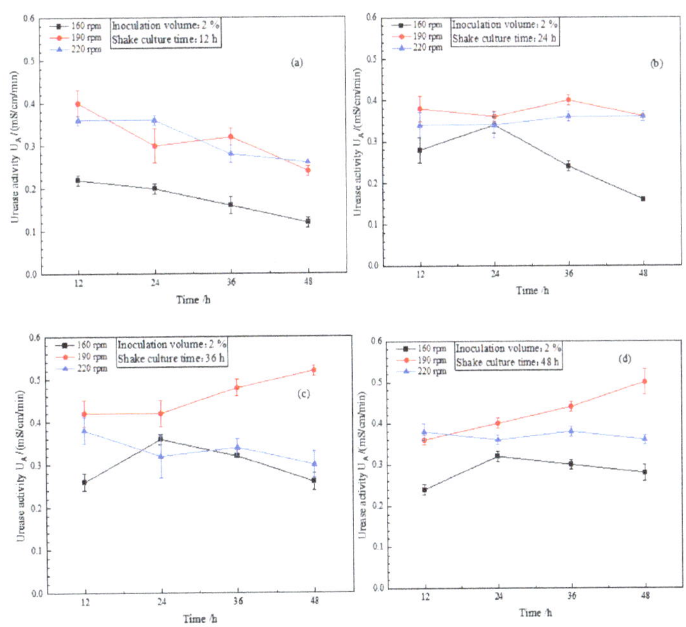

Figure 7. Urease activity after inoculation with 2% of the bacterial suspension with different shaking culture times at (**a**) 12 h; (**b**) 24 h; (**c**) 36 h; (**d**) 48 h.

Based on Figures 6–8, it can be concluded that, with the increase in inoculum, the pattern of change in urease activity of the bacterial bacterial suspension incubated at low speed was always virtually the same. On the other hand, the activity of the medium to high-speed shaking culture for 12 h gradually changed from stable to decaying after resting, and the activity of the shaking 24 h after resting became larger. After 36 h of shaking, the activity of the medium-speed culture was significantly higher than the activity of the other two groups in both the shaking period and the resting period, and the inoculum level had only a minor effect on the urease activity of the high- and low-speed shaking cultures.

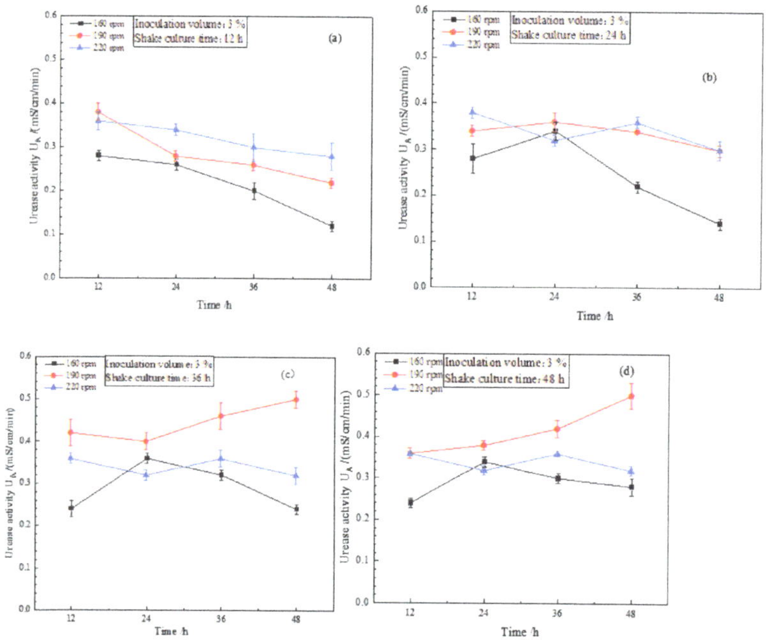

Figure 8. Urease activity after inoculation with 3% of the bacterial suspension with different shaking culture times at (**a**) 12 h; (**b**) 24 h; (**c**) 36 h; (**d**) 48 h.

3.1.3. Effect of Shake Culture Time on Urease Activity

In Figure 9, the urease activity is depicted at different times after inoculation with a range of bacterial suspension concentrations and after expansion of the culture at multiple speeds. The urease activity values of all groups increased significantly from 12 to 24 h in the shaking culture, with a peak value of 0.36 mS/cm/min. After 24 h, the urease activity of all the bacteria began to decay, and the shorter the shaking culture time, the faster the decay rate, and the longer the shaking culture time, the higher the final urease activity value. Even if the culture was shaken for only 12 h and left to rest, the urease activity decayed continuously, and with the increase in the inoculum, the final urease activity value decayed to 0.12 mS/cm/min, which is about 1/3 of the peak value. It is recommended to use it immediately following a peak in urease activity at 24 h.

As shown in Figure 10, urease activity varies with time when incubated at varying rotational speeds. The urease activity increased with an increase in rotation speed across all time periods. In terms of bacterial activity, the bacteria cultured at a high speed (220 rpm) remained active, the bacteria cultured at a medium speed (190 rpm) decreased and then stabilized, and the bacteria cultured at a low speed (160 rpm) continued to decay. After two days of incubation, the urease activity of all groups progressively increased with increasing shaking time. While the urease activity in both medium- and high-speed cultures was more stable and the medium-speed culture was slightly better, the urease activity in the low-speed shaking culture began to decline after reaching its peak at 24 h, and the activity decayed more quickly under rest.

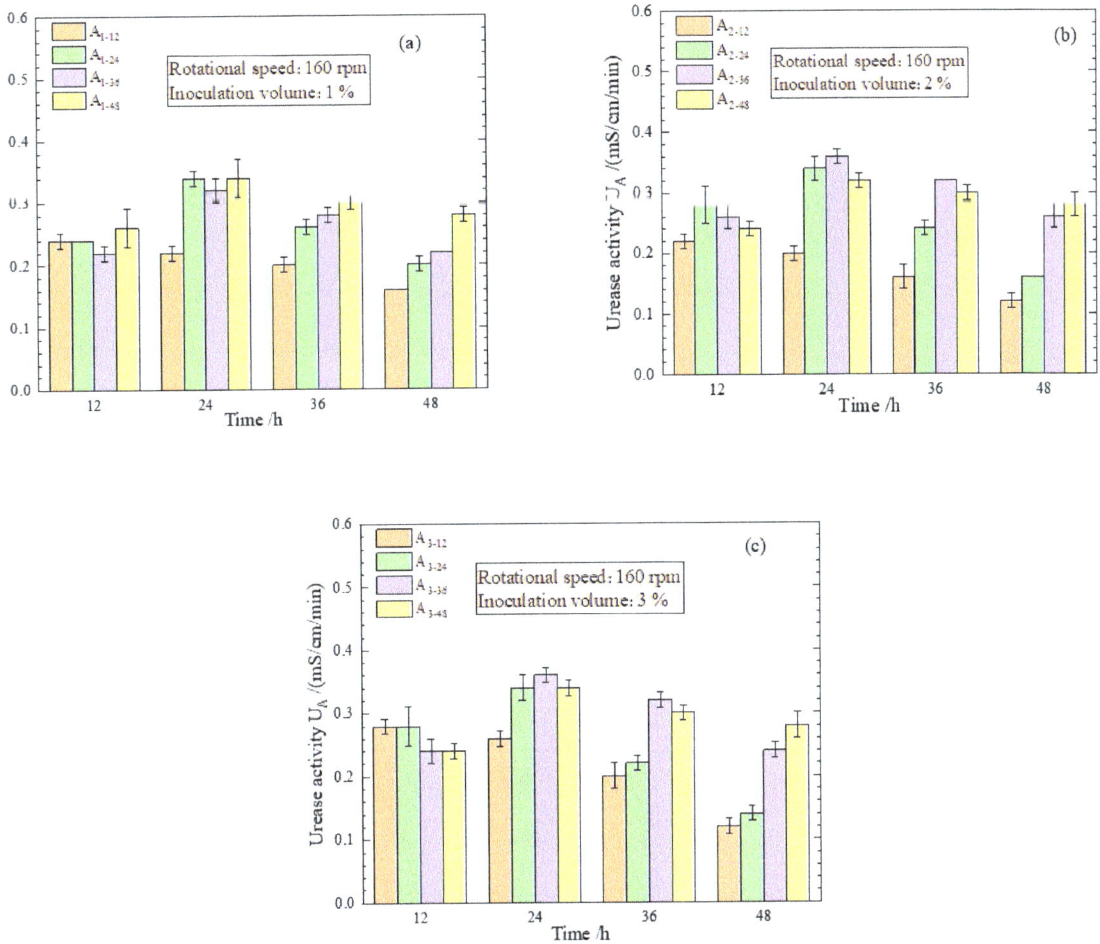

Figure 9. Urease activity at different moments of expansion with 160 rpm rotation speed after inoculation with (**a**) 1%; (**b**) 2%; (**c**) 3% bacterial suspension.

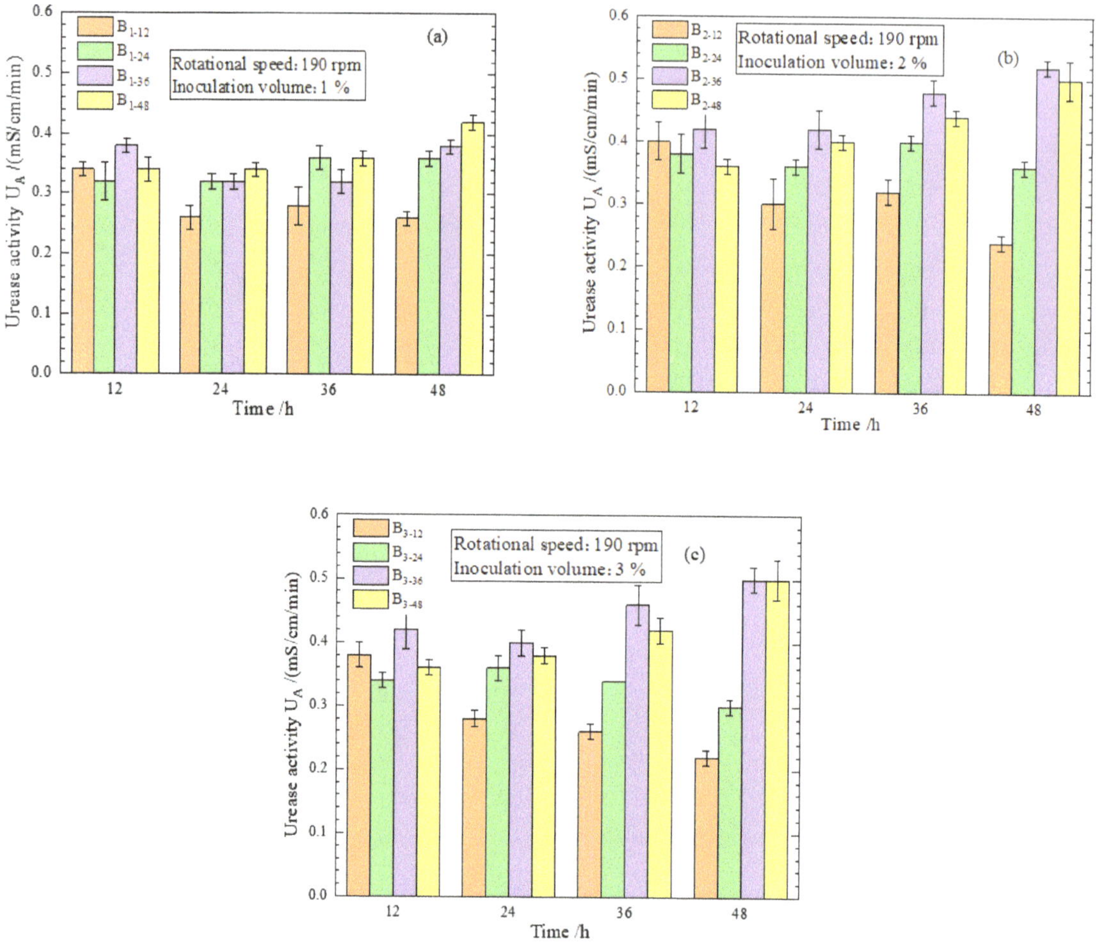

Figure 10. Urease activity at different moments of expansion with 190 rpm rotation speed after inoculation with (**a**) 1%; (**b**) 2%; (**c**) 3% bacterial suspension.

The results in Figure 11 show that when incubation was carried out for only 12 h, the urease activity varied with the inoculum level, with the low inoculum level maintaining the activity for a longer period. In contrast, the medium and high inoculum levels declined after 12 h. When shaking time reached 24 h and above, the urease activity values and the changing pattern were almost similar for all groups at all time periods, and the urease activity values were always maintained within a certain activity range. When the rotation speed is high, using a low inoculum level and a short shaking time is recommended.

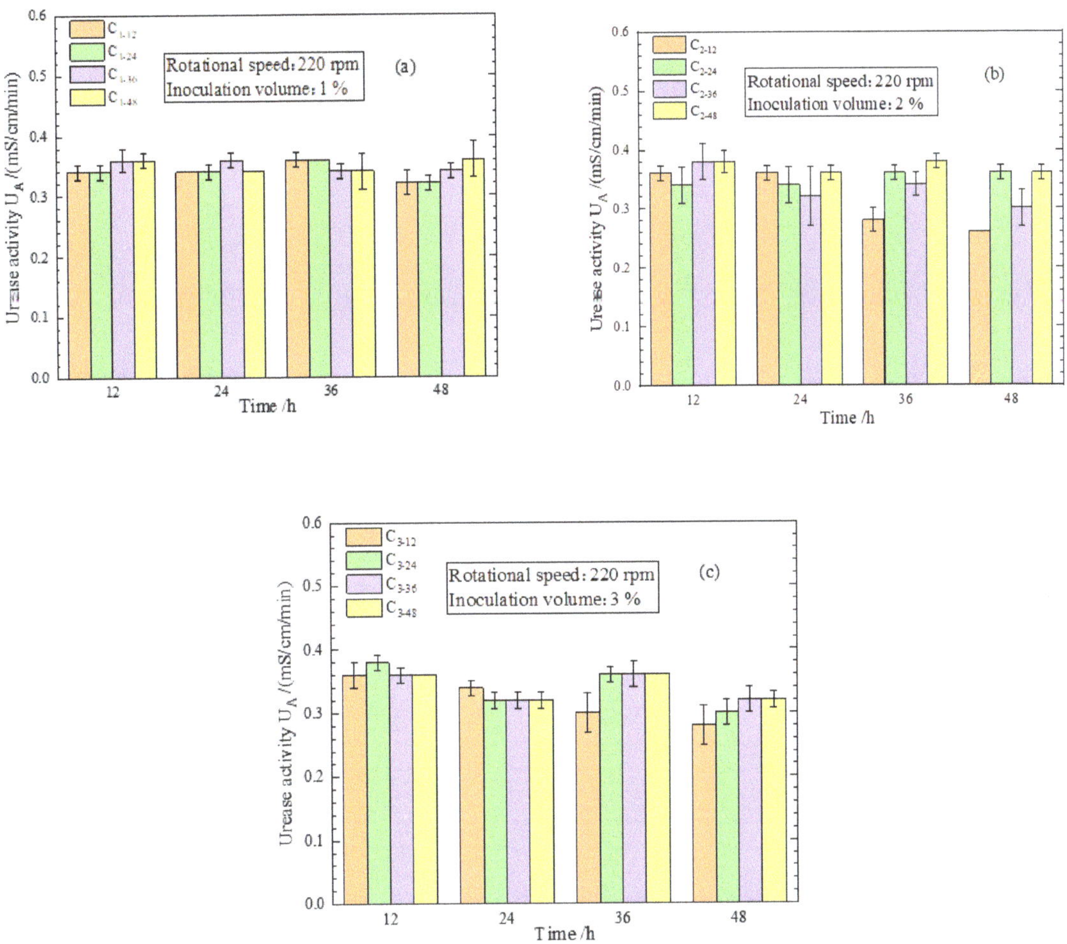

Figure 11. Urease activity at different moments of expansion with 220 rpm rotation speed after inoculation with (**a**) 1%; (**b**) 2%; (**c**) 3% bacterial suspension.

3.2. Discussion of Peak Urease Activity

To replenish the bacterial suspension in time and to avoid prolonging the construction process, a bubble map of the peak urease activity and the corresponding program parameters can be used to produce a bacterial suspension with the target urease level. By replacing scattered points with individual bubbles, the bubble map represents three-dimensional information instead of a scattered point map. As shown in Figure 12, the X axis represents the volume of inoculum, the Y axis represents time, and the Z axis represents the size of the bubbles. In addition, the color mapping allows visualization of the peak activity that a culture of bacteria can achieve under specific conditions.

It was observed that the high-activity-value bubbles were mainly distributed in the region of inoculum $\geq$ 2% and shaking incubation time $\geq$ 36 h. Under the conditions of suitable inoculum level, the peak urease activity could be greatly increased after 24 h after continued incubation, while when the inoculum level was too low; the effect after 48 h of incubation was almost identical to that of 12 h of incubation under optimum conditions. In this experiment, the urease activity values of the bacterial suspension prepared using

Scheme B2-36 were the highest at all time points. Table 3 lists the experimental conditions when the activity of the bacterial solution reached its maximum value.

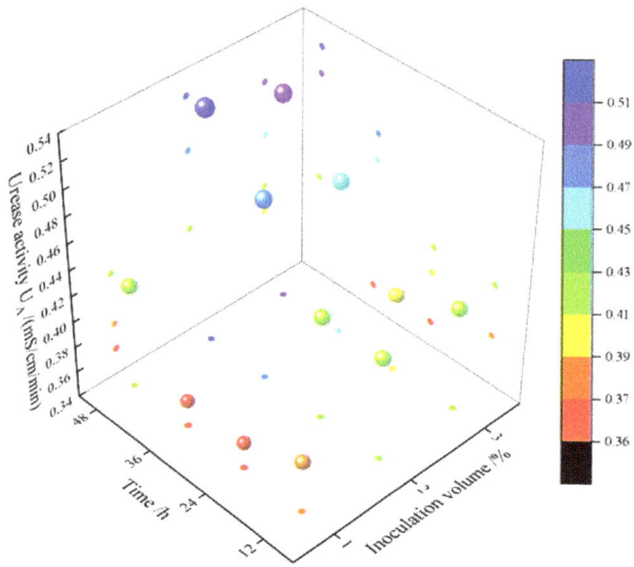

Figure 12. Bubble map of urease activity peak.

Table 3. Urease activity peak and scheme.

	12 h	24 h	36 h	48 h
1%	0.38	0.36	0.36	0.42
Number	B_{1-36}	C_{1-36}	B_{1-24}, B_{1-48}, C_{1-12}, C_{1-24}	B_{1-48}
2%	0.42	0.42	0.48	0.52
Number	B_{2-36}	B_{2-36}	B_{2-36}	B_{2-36}
3%	0.42	0.40	0.46	0.50
Number	B_{3-36}	B_{3-36}	B_{3-36}	B_{3-36}, B_{3-48}

It was found that incubating the bacterial suspension at low rotational speed had a relatively small effect on the peak of urease activity, and the trend in activity change and the time point of peak activity tended to be similar. Nonetheless, as the speed of shaking incubation was increased and the incubation time was extended, there were significant differences in the peak activity and the pattern of change in the bacterial suspension as a function of the inoculum level. An inoculum level of 2% was found to be most effective.

The pattern of urease activity change between the two protocols differed as the shaker speed was varied. Generally, the urease activity levels were low for the medium and high speeds, whereas the values for the low speed were higher. The peak urease activity of the medium-speed condition was much higher than that of the other two conditions in the optimum environment.

The effect of prolonging the shaking time on the urease activity of the broth was mainly in terms of "slowing down the decay of activities" and "promoting the increase in activities". The continued shaking of the bacterial suspension will promote the maintenance of urease activity if the urease activity of the broth has reached its peak; if the urease activity of the broth has not reached its peak, extending the shaking time will promote the increase in urease activity.

Following 36 h of shaking at the optimal inoculum and speed, the urease activity values significantly increased both in shaking and standing cultures, and for economic reasons, the standing culture should be used. At all test time points, the highest levels of urease activity were measured for the solution prepared in accordance with protocol $B_{2\text{-}36}$, which involves inoculation of 2% bacterial suspension under 190 rpm rotation speed and shaking for 36 h.

Author Contributions: Conceptualization, J.H.; methodology, J.H.; software, Y.W.; validation, Y.Z.; formal analysis, Y.Z.; investigation, D.Z.; resources, H.Z.; data curation, Z.W.; writing—original draft preparation, Y.W.; writing—review and editing, Y.W.; visualization, Y.W.; supervision, J.H.; project administration, J.H.; funding acquisition, H.Z. and D.Z. All authors have read and agreed to the published version of the manuscript.

Funding: This research was funded by the National Natural Science Foundation of China (51968019), the Hainan Provincial Natural Science Foundation Innovation Research Team Project (522CXTD511), the High Technology Direction Project of the Key Research & Development Science and Technology of Hainan Province, China (ZDYF2021GXJS020), the High-Level Talent Project of Hainan Basic and Applied Basic Research Plan (2019RC148, 2019RC351), and the characteristic innovation (Natural Science) projects of scientific research platforms and scientific research projects of Guangdong Universities in 2021 (2021KTSCX139). The authors also acknowledge support from the China Scholarship Council (201907565040).

Institutional Review Board Statement: Not applicable.

Informed Consent Statement: Not applicable.

Data Availability Statement: Not applicable.

Conflicts of Interest: The authors declare no conflict of interest.

References

1. Yang, S.; Shen, X.; Liu, H.; Ge, H.; Rui, X. Gradation affects basic mechanical characteristics of Chinese calcareous sand as airport subgrade of reefs. *Mar. Georesour. Geotechnol.* **2020**, *38*, 706–715. [CrossRef]
2. Wang, X.; Liu, J.Q.; Cui, J.; Wang, X.Z.; Shen, J.H.; Zhu, C.Q. Particle breakage characteristics of a foundation filling material on island-reefs in the South China Sea. *Constr. Build. Mater.* **2021**, *306*, 124690. [CrossRef]
3. Cheng, L.; Cord-Ruwisch, R.; Shahin, M.A. Cementation of sand soil by microbially induced calcite precipitation at various degrees of saturation. *Can. Geotech. J.* **2013**, *50*, 81–90. [CrossRef]
4. Han, Z.; Cheng, X.; Ma, Q. An experimental study on dynamic response for MICP strengthening liquefiable sands. *Earthq. Eng. Eng. Vib.* **2016**, *15*, 673–679. [CrossRef]
5. Lin, H.; Suleiman, M.T.; Brown, D.G. Investigation of pore-scale $CaCO_3$ distributions and their effects on stiffness and permeability of sands treated by microbially induced carbonate precipitation (MICP). *Soils Found.* **2020**, *60*, 944–961. [CrossRef]
6. Whitaker, J.M.; Vanapalli, S.; Fortin, D. Improving the strength of sandy soils via ureolytic $CaCO_3$ solidification by Sporosarcina ureae. *Biogeosciences* **2018**, *15*, 4367–4380. [CrossRef]
7. Liu, S.; Yu, J.; Zeng, W.; Peng, X.; Cai, Y.; Tu, B. Study on the effect of microbially induced calcium carbonate precipitation on repairing cracks in triangular soils. *J. Rock Mech. Eng.* **2020**, *39*, 191–204.
8. Yue, J.W.; Chen, Y.; Zhao, L.M.; Zhang, B.; Kong, Q.M.; Gu, L.H.; Lu, H.F. Strength characteristics and mechanism of action of glutinous rice slurry improved imitation site soil at Chi Cheng site. *J. Civ. Environ. Eng.* **2022**, *44*, 195–204. (In Chinese and English)
9. Mu, B.; Gui, Z.; Lu, F.; Petropoulos, E.; Yu, Y. Microbial-Induced Carbonate Precipitation Improves Physical and Structural Properties of Nanjing Ancient City Walls. *Materials* **2021**, *14*, 5665. (In Chinese) [CrossRef] [PubMed]
10. Liu, S.; Yu, J.; Peng, X.; Cai, Y.; Tu, B. Preliminary study on repairing tabia cracks by using microbially induced carbonate precipitation. *Constr. Build. Mater.* **2020**, *248*, 118611. (In Chinese) [CrossRef]
11. Cheng, L.; Shahin, M.A. Stabilisation of oil-contaminated soils using microbially induced calcite crystals by bacterial flocs. *Géotechnique Lett.* **2017**, *7*, 146–151. [CrossRef]
12. Chen, X.; Zhang, D.; Larson, S.L.; Ballard, J.H.; Knotek-Smith, H.M.; Nie, J.; Hu, N.; Ding, D.; Han, F.X. Microbially Induced Carbonate Precipitation Techniques for the Remediation of Heavy Metal and Trace Element–Polluted Soils and Water. *Water Air Soil Pollut.* **2021**, *232*, 268. [CrossRef]
13. Wei, T.; Yashir, N.; An, F.; Imtiaz, S.A.; Li, X.; Li, H. Study on the performance of carbonate-mineralized bacteria combined with eggshell for immobilizing Pb and Cd in water and soil. *Environ. Sci. Pollut. Res.* **2022**, *29*, 2924–2935. [CrossRef] [PubMed]
14. Lyu, C.; Qin, Y.; Chen, T.; Zhao, Z.; Liu, X. Microbial induced carbonate precipitation contributes to the fates of Cd and Se in Cd-contaminated seleniferous soils. *J. Hazard. Mater.* **2022**, *423*, 126977. [CrossRef] [PubMed]

15. Rahman, M.M.; Hora, R.N.; Ahenkorah, I.; Beecham, S.; Karim, M.R.; Iqbal, A. State-of-the-Art Review of Microbial-Induced Calcite Precipitation and Its Sustainability in Engineering Applications. *Sustainability* **2020**, *12*, 6281. [CrossRef]
16. Stocks-Fischer, S.; Galinat, J.K.; Bang, S.S. Microbiological precipitation of $CaCO_3$. *Soil Biol. Biochem.* **1999**, *31*, 1563–1571. [CrossRef]
17. De Muynck, W.; De Belie, N.; Verstraete, W. Microbial carbonate precipitation in construction materials: A review. *Ecol. Eng.* **2010**, *36*, 118–136. [CrossRef]
18. Fujita, M.; Nakashima, K.; Achal, V.; Kawasaki, S. Whole-cell evaluation of urease activity of Pararhodobacter sp. isolated from peripheral beachrock. *Biochem. Eng. J.* **2017**, *124*, 1–5. [CrossRef]
19. Zhao, X. Experimental Study of Microbially Induced Calcium Carbonate Precipitation (MICP) Curing Soil. Ph.D. Thesis, China University of Geosciences, Beijing, China, 2014. Available online: https://cdmd.cnki.com.cn/Article/CDMD-11415-1014249450.htm (accessed on 14 June 2022). (In Chinese)
20. Lv, C.; Tang, C.; Zhu, C.; Li, W.; Chen, T.; Zhao, L.; Pan, X. Environmental Dependence of Microbially Induced Calcium Carbonate Crystal Precipitations: Experimental Evidence and Insights. *J. Geotech. Geoenviron. Eng.* **2022**, *148*, 04022050. [CrossRef]
21. Liang, S.; Xiao, X.; Li, Z.; Feng, D. Effect of Nutrient Solution Composition on Bio-Cemented Sand. *Crystals* **2021**, *11*, 1572. [CrossRef]
22. Tang, C.; Yin, L.; Jiang, N.; Zhu, C.; Zeng, H.; Li, H.; Shi, B. Factors affecting the performance of microbial-induced carbonate precipitation (MICP) treated soil: A review. *Environ. Earth Sci.* **2020**, *79*, 94. [CrossRef]
23. Al-Thawadi, S. High Strength In-Situ Biocementation of Soil by Calcite Precipitating Locally Isolated Ureolytic Bacteria. Ph.D. Thesis, Murdoch University, Perth, Australia, 2008. Available online: https://researchrepository.murdoch.edu.au/id/eprint/721/ (accessed on 14 June 2022).
24. Lai, H.J.; Cui, M.J.; Wu, S.F.; Yang, Y.; Chu, J. Retarding effect of concentration of cementation solution on biocementation of soil. *Acta Geotech.* **2021**, *16*, 1457–1472. [CrossRef]
25. Han, Z.G.; Cheng, X.H. Exploration of the effect of nutrient salts on microbial reinforcement of liquefiable sandy soils. *Ind. Constr.* **2015**, *45*, 19–22. (In Chinese)
26. Zhang, Y.; Guo, H.X.; Cheng, X.H. Influences of calcium sources on microbially induced carbonate precipitation in porous media. *Mater. Res. Innov.* **2014**, *18* (Suppl. S2), S2-79—S2-84. [CrossRef]
27. Li, C.; Wei, T.; Ji, B.; Lei, X.; Wang, X. Effects of different calcium sources and Ca^{2+} concentrations on MICP. *Environ. Sci. Technol.* **2018**, *41*, 30–34. (In Chinese)
28. Maleki Kakelar, M.; Yavari, M.; Yousefi, M.R.; Nimtaj, A. The Influential Factors in the Effectiveness of Microbial Induced Carbonate Precipitation (MICP) for Soil Consolidation. *J. Hum. Environ. Health Promot.* **2020**, *6*, 40–46. [CrossRef]
29. Erşan, Y.Ç.; de Belie, N.; Boon, N. Microbially induced $CaCO_3$ precipitation through denitrification: An optimization study in minimal nutrient environment. *Biochem. Eng. J.* **2015**, *101*, 108–118. [CrossRef]
30. Wang, Y.; Jin, Z.; Wu, L. Preservation and management of microbial strains. *Shanghai Prev. Med.* **2007**, *19*, 87–88.
31. Chu, J.; Stabnikov, V.; Ivanov, V. Microbially Induced Calcium Carbonate Precipitation on Surface or in the Bulk of Soil. *Geomicrobiol. J.* **2012**, *29*, 544–549. (In Chinese) [CrossRef]
32. Chen, H. Study on the Influence of Compression and Shear Properties of Grade Paired MICP Reinforced Sandy Soils. Master's Thesis, Chongqing University, Chongqing, China, 2019. Available online: https://cdmd.cnki.com.cn/Article/CDMD-10611-1019905958.htm (accessed on 14 June 2022).
33. Xu, W.; Zheng, J.; Chu, J.; Zhang, R.; Cui, M.; Lai, H.; Zeng, C. New method for using N-(N-butyl)-thiophosphoric triamide to improve the effect of microbial induced carbonate precipitation. *Constr. Build. Mater.* **2021**, *313*, 125490. (In Chinese) [CrossRef]
34. Whiffin, V.S. Microbial $CaCO_3$ Precipitation for the Production of Biocement. Ph.D. Thesis, Murdoch University, Perth, Australia, 2004. Available online: https://researchrepository.murdoch.edu.au/id/eprint/399/ (accessed on 14 June 2022).

sustainability

MDPI

Article

Experimental Study on Stiffness Degradation of Organic Matter-Disseminated Sand under Cyclic Loading

Juan Du [1,2,*], Xingfei Jiang [1], Bingyang Liu [3], Lin Jia [4] and Yang Zhang [1,*]

1 School of Civil Engineering and Architecture, Hainan University, Haikou 570228, China
2 School of Civil Engineering, Tianjin University, Tianjin 300072, China
3 School of Civil Engineering, Harbin Institute of Technology, Harbin 150006, China
4 Hainan Investigation Institute of Hydrogeology and Engineering Geology, Haikou 570206, China
* Correspondence: dujuan2012@hainanu.edu.cn (J.D.); a314261366@163.com (Y.Z.)

Abstract: Degradation of stiffness will occur in sand under cyclic loading. Organic matter-disseminated sand (OMDS) is a special sand in the northeast and western coastal region of Hainan Island. Through the stress-controlled dynamic triaxial test, the natural moisture content, vibration amplitude and consolidation ratio of OMDS under three types of cyclic loads (sine wave, triangular wave and rectangular wave) were studied. The results showed that the soil stiffness decreases with the increase in vibration times. The increase in natural moisture content and vibration amplitude, and the reduction in the consolidation ratio accelerate the softening of soil stiffness. Furthermore, based on the test results, an empirical formula was derived to reflect the rule of soil stiffness softening.

Keywords: organic matter-disseminated sand (OMDS); dynamic deformation; stiffness degradation; softening index

1. Introduction

With the construction of the Hainan free trade port, many structures are to be built in coastal areas. Organic matter-disseminated sand (OMDS) is a special sand in the northeast and western coastal region of Hainan Island, shown in Figure 1a. Figure 1b shows the scanning electron microscope photos of OMDS. It can be seen that the surface of the OMDS particles are obviously smoother, and there are fewer pores in the surface, which indicates that the organic matter has been immersed in pores and adsorbed to the surface of the sand particles, forming a film that wraps around the sand particles. The influence of waves, traffic and earthquakes on OMDS's foundation should be of concern in engineering. In a wide range of studies, traffic loads could be approximately equivalent to triangular waves and sine waves, and wave loads could be equivalent to rectangular waves [1]. With the increase in dynamic cyclic load amplitude, the dynamic strain, porosity and pore pressure of soil increase, the internal structure of soil is reshaped as well, which induces the stiffness degradation of soil [2–4]. Under the cyclic load, the cumulative deformation of soil varies with vibration cycles [5,6], and due to the nonlinearity of soil, the stiffness acts in a complicated manner [7,8].

The characteristics of the long-term deformation and stiffness softening of soil under cyclic loading has been researched extensively by various countries' scholars. Monismith et al. [9] proposed an exponential model for the relationship between the plastic strain of soft clay and vibration cycles. Introducing dynamic stress and static strength, Li and Selig [10,11] revised the Monismith function model and established a new function model. Idriss et al. [12] analyzed the relationship between the degradation index and vibration cycles, and established a functional relationship to describe the quantitative attenuation rule of soil cyclic stiffness. Yao et al. [13] modified the Idriss formula and redefined the softening index to describe the stiffness degradation of soft clay. Chai et al. [14] and Parr et al. [15] studied the dynamic characteristics of clay through dynamic triaxial tests, and obtained the double logarithmic relationship between dynamic strain and vibration cycles. Huurman [16] suggested that the stiffness degradation of sand was related to the cyclic stress ratio and vibration cycles. Yasuhara et al. [17] proposed a modified model function describing the relationship of the degradation index and vibration cycles based on the dynamic triaxial test of soil under cyclic loading. Tan et al. [18] analyzed the results of the cyclic triaxial test and cyclic simple shear test of cohesive soil with different plasticity indexes. Wang et al. [19] established a functional relationship between the degradation index and vibration cycles, and obtained the fracture toughness ratio of clay through the function. Huang et al. [20] introduced the dynamic deviatoric stress parameter into a functional model describing the relationship of the degradation index and vibration cycles. Cao et al. [21] established a stiffness degradation function for marine soft soil, and the relationship of the degradation index and vibration cycles was obtained by the function. Hong et al. [22] carried out a series of stress-path triaxial tests, with local strain measurements, to achieve path-dependent stiffness degradation curves of medium-dense Toyoura sand. Iraji et al. [23] modified the nonlinear Pastor–Zienkiewicz–Chan constitutive model to simulate the stiffness degradation of dense sands at dynamic loading. Mei et al. [24] analyzed the development laws between the axial permanent strain of coarse-grained soil and vibration cycles. Liu et al. [25] found that the softening index was exponentially related to cumulative pore pressure and cumulative strain.

In summary, it can be seen that the stiffness softening of clay, silt and sand under cyclic loading has been studied, while the dynamic properties of sand with organic matter are different from those of general sand. There are few reports on its stiffness degradation. In this study, the degradation laws of skeleton curves on OMDS samples were studied under three cyclic loadings with various natural moisture contents, consolidation ratios, confining pressures and vibration amplitudes.

(a)

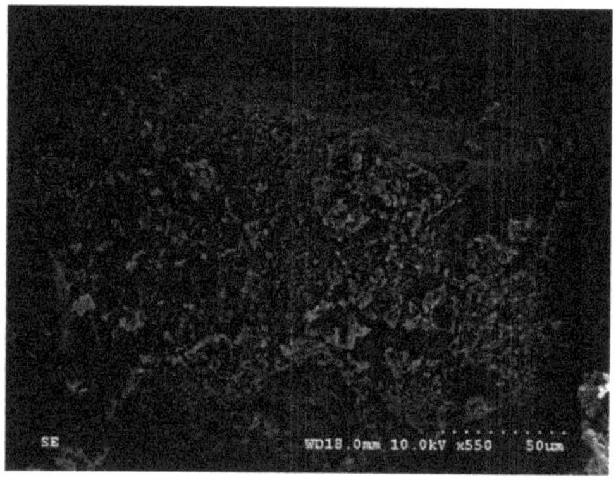

(b)

Figure 1. OMDS photos: (**a**) on-site photo; (**b**) SEM photo.

2. Materials and Methods

2.1. Materials

The OMDS [26] was obtained from a site in Hainan Province, the upper layer soil to 1.2 m depth was cultivated soil, and the lower layer was OMDS, the sand used in this experimental study was extracted at depth ranging from 2 m to 5 m by thin-walled tube sampler. The coefficient of uniformity C_u was 2.07, the coefficient of curvature C_c was 1.25, the main physical parameters and particle grading curves are shown in Table 1 and Figure 2, respectively.

Table 1. Main physical parameters of OMDS.

Organic Content/%	Natural Moisture Content/%	Natural Density/g·cm^{-3}	Minimum dry Density/g·cm^{-3}	Maximum Dry Density/g·cm^{-3}	Specific Gravity
5.32	12.23	1.617	1.570	1.723	2.62

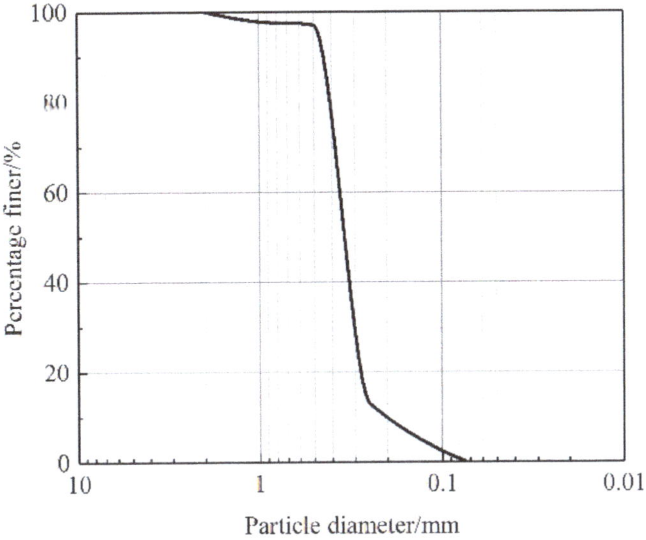

Figure 2. Grading curve of OMDS.

2.2. Sample Preparation

The OMDS samples were first air-dried and then passed through a sieve with opening size of 5 mm. Fully mixed with water and placed for 24 h, the sieved soil was cast into cylindrical molds which had 39.1 mm diameter and 80 mm height, as shown in Figure 3. First, the pumping saturation method was used for saturation, and then the back-pressure saturation was carried out by the dynamic triaxial instrument, until the sample saturation reached 95%. Lastly, the fully saturated samples were consolidated under the condition of isotropic consolidation.

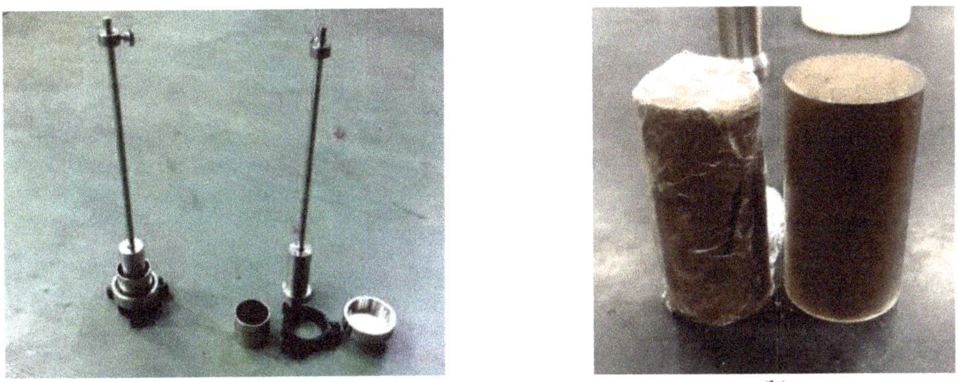

(**a**) (**b**)

Figure 3. Compacted OMDS samples: (**a**) compaction tool; (**b**) OMDS samples.

2.3. Scheme of Dynamic Triaxial Test

In this study, the dynamic responses of OMDS under multiple cyclic loading waves were studied, and the influence of vibration amplitude and vibration cycles was considered as well; therefore, DDS-70 dynamic triaxial test equipment shown in Figure 4 was adopted. The DDS-70 system can automatically generate sine waves, triangular waves and rectangular waves with various frequencies, the maximum frequency is 10 Hz. The confining pressure of the test was 100 kPa. Because the frequency had little influence on the sand sample [27], the test frequency was set 1.0, and the test process was undrained. The test scheme is shown in Table 2.

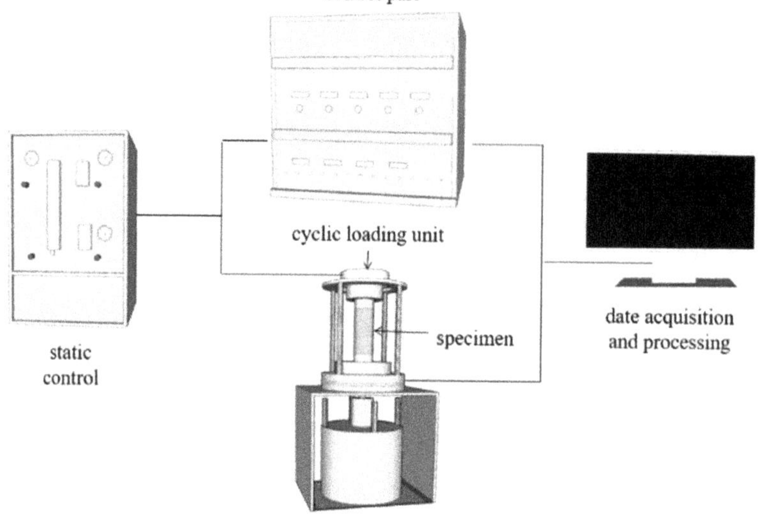

Figure 4. Schematic drawing of DDS-70.

Table 2. Scheme of dynamic triaxial test.

Sample	Consolidation Ratio	Natural Moisture Content/%	Wave Form	Vibrating Load/N
OMDS	1.0	10%	sine, triangular, rectangular	25, 70, 110, 150
		18%	sine, triangular, rectangular	25, 70, 110, 150
		24%	sine, triangular, rectangular	25, 70, 110, 150
	1.5	10%	sine, triangular, rectangular	25, 70, 110, 150
		18%	sine, triangular, rectangular	25, 70, 110, 150
		24%	sine, triangular, rectangular	25, 70, 110, 150
	2.0	10%	sine, triangular, rectangular	25, 70, 110, 150
		18%	sine, triangular, rectangular	25, 70, 110, 150
		24%	sine, triangular, rectangular	25, 70, 110, 150

3. Results

3.1. Skeleton Curve

The skeleton curve demonstrates the relationship between the dynamic shear stress (τ_d) and the dynamic shear strain (γ_d) [28,29], which could be expressed in terms of the dynamic axial stress (σ_d) and the dynamic axial strain (ε_d), as shown in Equations (1) and (2) [30,31].

$$\tau_d = 0.5\sigma_d \tag{1}$$

$$\gamma_d = \varepsilon_d(1 + \mu) \tag{2}$$

where τ_d is the dynamic shear stress; σ_d is the dynamic axial stress; γ_d is the dynamic shear strain; ε_d is the dynamic axial strain; and μ is the Poisson's ratio. Since the lateral displacement of the samples were not measured in this study, it was not possible to experimentally determine the Poisson's ratio value. Therefore, instead of using τ_d and γ_d, σ_d and ε_d were used to construct the skeleton curve.

Figure 5 shows the skeleton curves of ODMS samples with varying moisture contents and consolidation ratios, and under four confining pressure levels when the wave form was a sine wave. It can be seen that, with the increase in confining pressure, σ_d became larger at an identical ε_d. Moreover, a lower moisture content and larger consolidation ratio in general resulted in the increase in σ_d at an identical ε_d. The above variation law between dynamic stress and dynamic strain under various wave forms was basically the same.

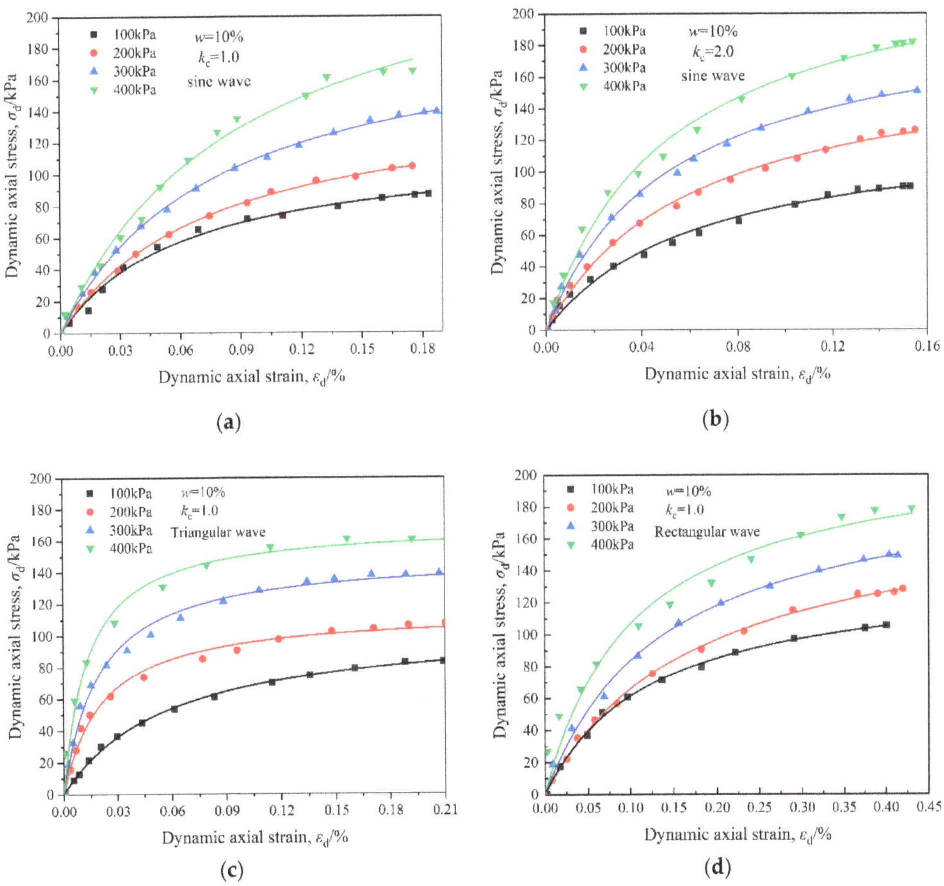

Figure 5. *Cont.*

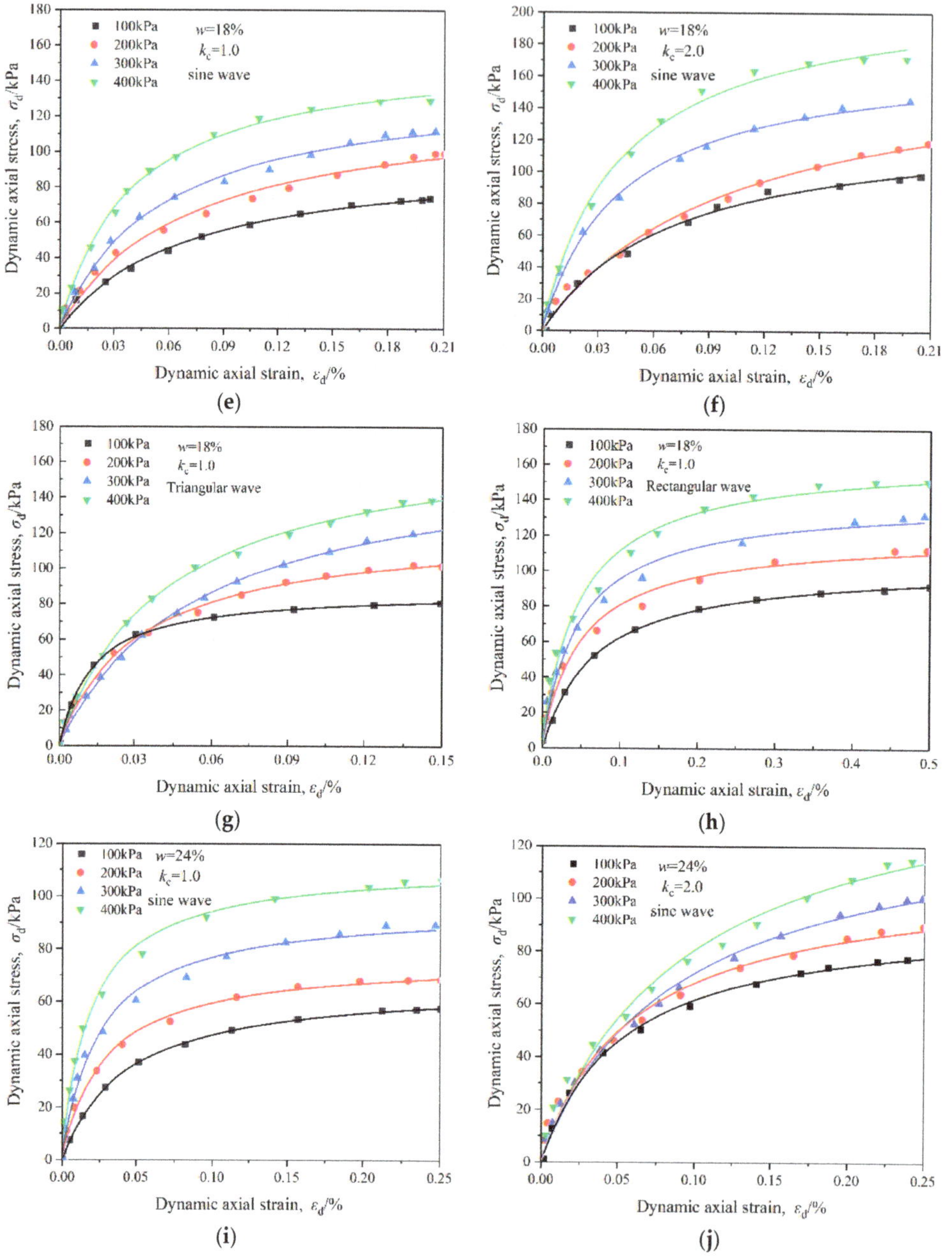

Figure 5. *Cont.*

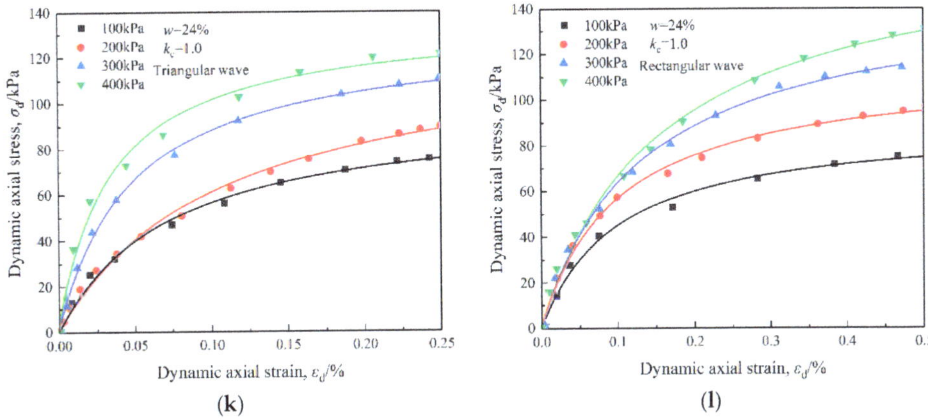

Figure 5. Skeleton curves of OMDS samples under dynamic triaxial test: (**a**) sine wave, 10% moisture content, 1.0 consolidation ratio; (**b**) sine wave, 10% moisture content, 2.0 consolidation ratio; (**c**) triangular wave, 10% moisture content, 1.0 consolidation ratio; (**d**) rectangular wave, 10% moisture content, 1.0 consolidation ratio; (**e**) sine wave, 18% moisture content, 1.0 consolidation ratio; (**f**) sine wave, 18% moisture content, 2.0 consolidation ratio; (**g**) triangular wave, 18% moisture content, 1.0 consolidation ratio; (**h**) rectangular wave, 18% moisture content, 1.0 consolidation ratio; (**i**) sine wave, 24% moisture content, 1.0 consolidation ratio; (**j**) sine wave, 24% moisture content, 2.0 consolidation ratio; (**k**) triangular wave, 24% moisture content, 1.0 consolidation ratio; (**l**) rectangular wave, 24% moisture content, 1.0 consolidation ratio.

3.2. Characteristics of Stiffness Degradation

Dynamic modulus is an important parameter of soil dynamic characteristics, which can reflect the stiffness characteristics under dynamic loads. Dynamic elastic modulus, E_d, is one dynamic modulus, which is defined as the ratio between σ_d and ε_d, plotted in Figure 6. It can be seen that, under conditions of the same moisture content and consolidation ratio, the various wave forms, E_d, became smaller at a given ε_d when the confining pressure was reducing.

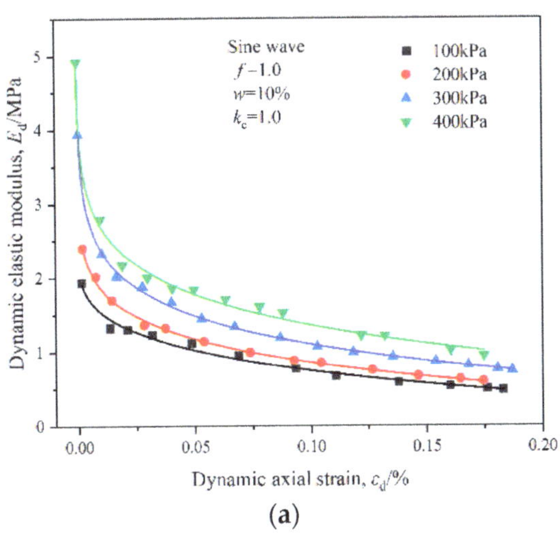

Figure 6. *Cont.*

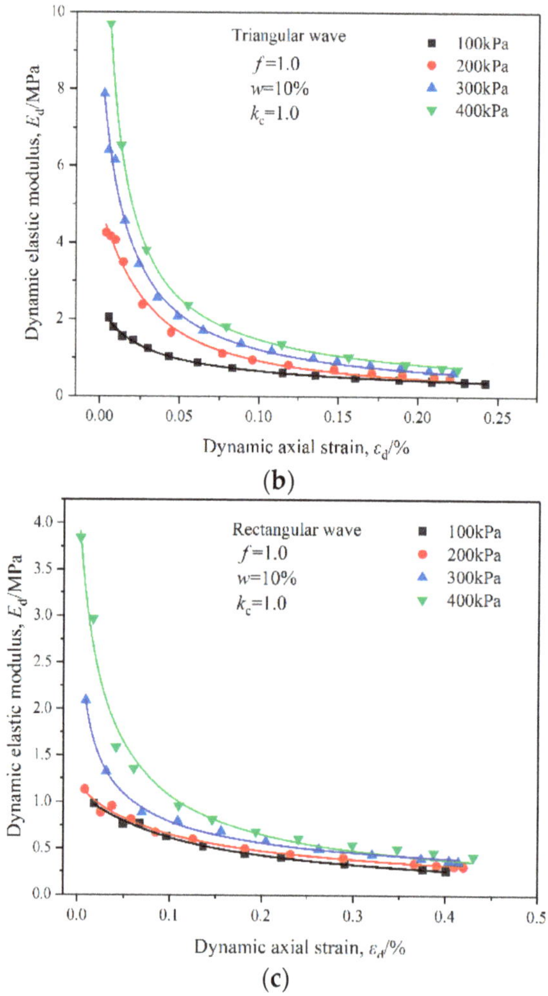

Figure 6. Variation in E_d and ε_d of OMDS samples under dynamic triaxial test: (**a**) sine wave; (**b**) triangular wave; (**c**) rectangular wave.

The degradation index, δ_d, can be used to quantify the attenuation of cyclic stiffness [8]. In the stress-controlled dynamic triaxial test, the ratio of the dynamic elastic modulus in the Nth cycle and the first cycle (i.e., the ratio of E_N:E_1) is the degradation index δ.

$$\delta_d = \frac{E_N}{E_1} = \frac{\sigma_d/(\varepsilon_d)_N}{\sigma_d/(\varepsilon_d)_1} = \frac{(\varepsilon_d)_1}{(\varepsilon_d)_N} \tag{3}$$

where $(\varepsilon_d)_N$ and $(\varepsilon_d)_1$ are the axial dynamic strain in the Nth and 1st cycle respectively, and σ_d is the dynamic axial stress.

3.2.1. Effects of Vibration Amplitude

Figure 7 shows the relationship of the degradation index and vibration cycles under various vibration amplitudes. It can be seen that the degradation index curve decreased sharply with the increase in vibration cycles under various cyclic loads, which indicates that the soil would be destroyed with fewer vibration cycles. Since the δ_d–N relationship is largely independent of the vibration amplitude, there was no critical stress for OMDS under cyclic loading, which was different from clay.

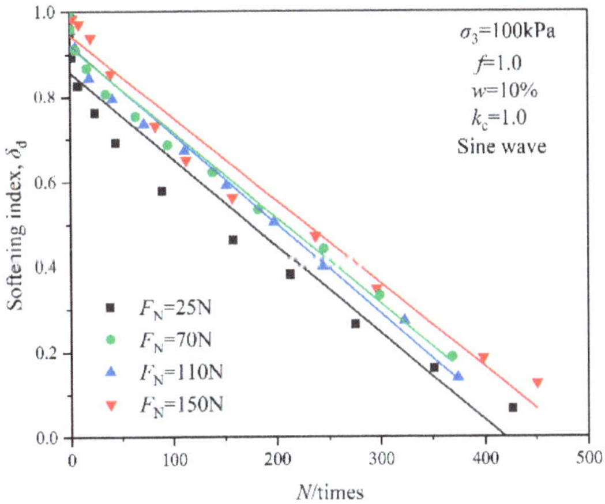

Figure 7. Relationship of degradation index and vibration cycles under various vibration amplitude.

3.2.2. Effects of Natural Moisture Content

Figure 8 shows the relationship of the degradation index and vibration cycles under various natural moisture contents. As observed, with the increase in vibration cycles, the degradation index curve tended to decrease rapidly in a nonlinear form, the variation range was smaller when the natural moisture content was smaller and the degradation degree was not obvious; when the natural moisture content reached 24%, the degradation index was larger than the other groups. This was consistent with the results reported by Zhong et al. [32] and was attributed to the fact that with the increase in moisture content, the water film surrounding the sand particles became thick, and the spacing between the sand particles increased. As a result, the interaction between the sand particles became weaker, which made the sand particles prone to dislocation deformation and the stiffness of OMDS prone to delegation. Figure 8 also shows that the curvatures of the degradation curves of various natural moisture contents were larger than those of vibration amplitude; the variation in natural moisture content had much influence on the degradation index.

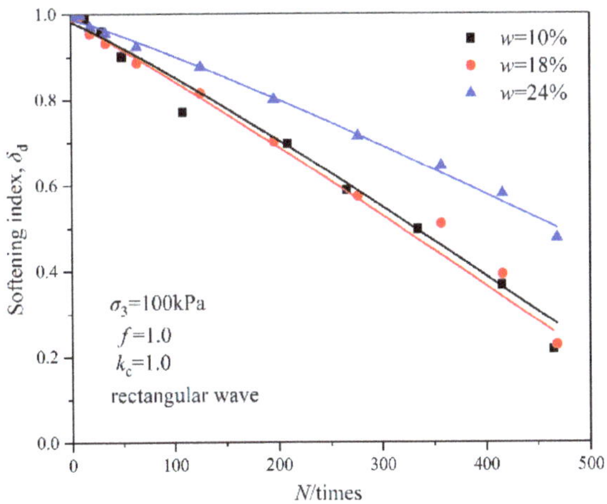

Figure 8. Relationship of degradation index and vibration cycles under various natural moisture contents.

3.2.3. Effects of Consolidation Ratio

Figure 9 shows the relationship of the degradation index and vibration cycles under various consolidation ratios. As observed, the degradation index curve decreased rapidly with the increase in vibration cycles. When the consolidation ratio k_c = 2.0, the variation rate of the degradation index curve was smaller than the other groups, which indicates that the pores' volume of soil decreased after consolidation, and the pores' water pressure also decreased in order to prevent the soil from instability or failure under vibration load. This was consistent with the results reported by Huang et al. [33] and was attributed to the fact that the larger the consolidation ratio, the greater the axial consolidation pressure, the higher the compaction degree of OMDS sample, the stronger the ability to resist elastic deformation, and the slower the stiffness degradation of OMDS. Figure 9 also shows that the curvature of the degradation index curve with different consolidation ratios was larger than that with different natural moisture contents, which shows a greater influence of the consolidation ratio on the degradation index.

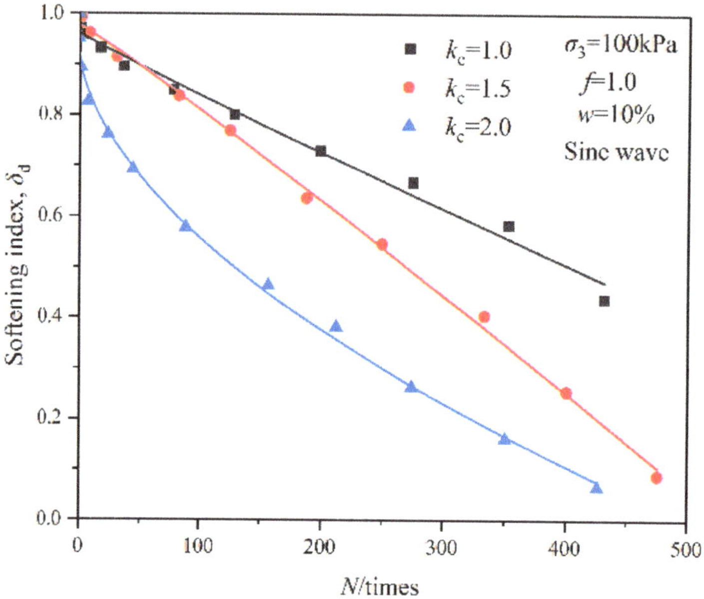

Figure 9. Relationship of degradation index and vibration cycles under various consolidation ratios.

3.2.4. Effects of Wave Form

Figure 10 shows the relationship of the degradation index and vibration cycles under various wave forms. As observed, the degradation index curves of OMDS showed nonlinear variation with the increase in vibration cycles under three kinds of wave loads. Under the same vibration cycles, the attenuation degree caused by the rectangular wave load was significantly greater than that by the sine wave and triangle wave. It was found that the soil was prone to degrade under the load of the rectangular wave. This was consistent with the results reported by Cao et al. [21] and was attributed to the fact that the work of the rectangular wave was much greater than that of the sine wave and triangle wave under the same vibration frequency, and the energy dissipation of soil was much greater as well.

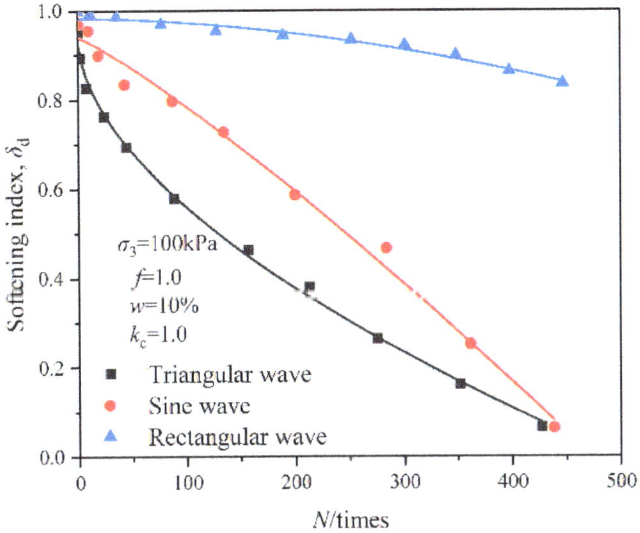

Figure 10. Relationship of degradation index and vibration cycles under various wave forms.

4. Degradation Model for OMDS

Based on the analysis of the test results and previous research, the functional relationship between the degradation index and vibration cycles of OMDS was established and is shown as

$$\delta_d = 1 + mN^n \quad (4)$$

where m and n are test parameters, obtained by regression analysis; N and δ_d are the vibration cycles and degradation index, respectively.

In this paper, Equation (4) was used to calculate the relationship between the degradation index and vibration cycles of OMDS under various influence factors, and the results were compared with those of dynamic triaxial test (see Figure 11).

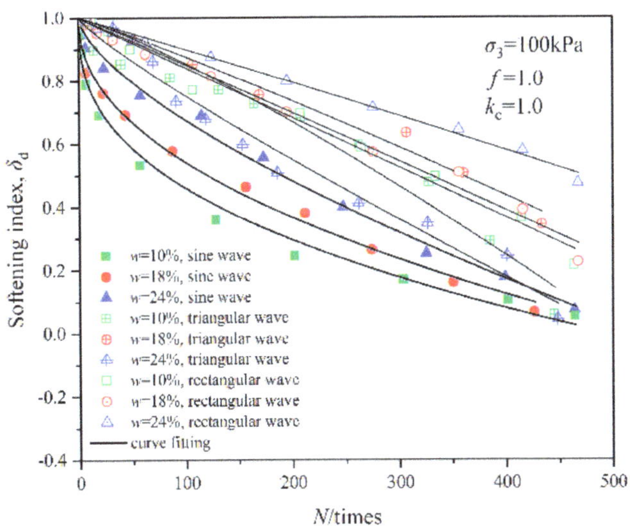

Figure 11. Fitting of degradation index under different cyclic loading wave forms.

The fitting results show that the application of Equation (4) can effectively simulate the softening curves of OMDS under cyclic loads with various amounts of moisture and wave forms such as the sine wave, triangular wave and rectangular wave, which can meet the needs of engineering construction (Table 3).

Table 3. Fitting parameters.

Wave Form	Moisture Content/%	Fitting Parameters		
		m	n	R^2
Sine wave	10	−0.01416	0.67934	0.87
	18	−0.05232	0.47059	0.90
	24	−0.09039	0.38753	0.89
Triangular wave	10	−0.00783	1.09822	0.79
	18	−0.00592	1.19321	0.82
	24	−0.00447	0.8703	0.85
Rectangular wave	10	−0.00837	1.03818	0.79
	18	−0.00134	1.02179	0.82
	24	−0.00157	1.00179	0.88

5. Discussion

Table 4 compares the degradation models constructed by several researchers, which were adopted to fit the degradation curve of OMDS with various moisture contents under sine wave cyclic loading, as shown in Figure 12 and Table 5.

Table 4. Degradation models.

Researcher	Degradation Model	Reference
Idriss et al.	$\delta_d = N^{-d}$	[12]
Yasuhara et al.	$\delta_d = 1 - d\lg N$	[17]
Yao et al.	$\delta_d = 10^{N^{-T}-1}$	[13]
This paper	$\delta_d = 1 + mN^n$	Present work

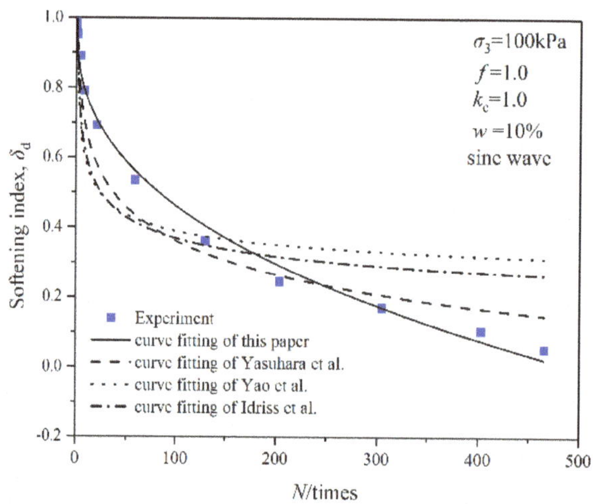

(a)

Figure 12. *Cont.*

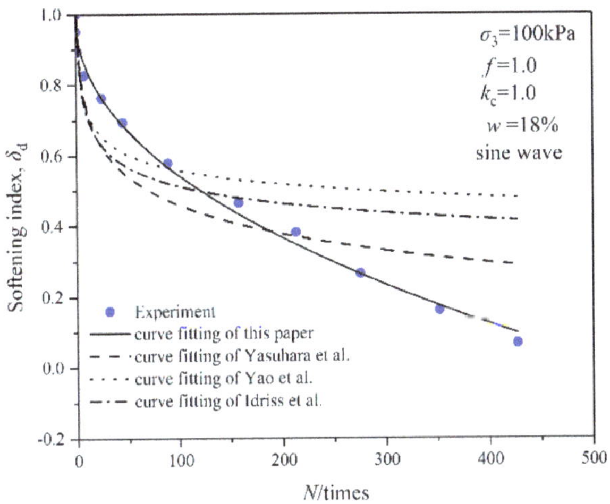

(b)

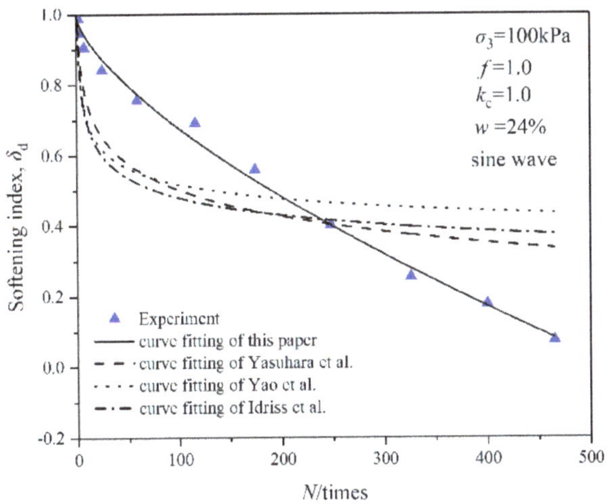

(c)

Figure 12. Comparison of degradation model: (**a**) w = 10%; (**b**) w = 18%; (**c**) w = 24%.

Table 5. Fitting effect of models based on the experimental data.

Sample	Wave Form	Moisture Content (%)	Residual Error Square Sum			
			Idriss Model	Yasuhara Model	Yao Model	This Paper
OMDS	Sine wave	10	1.198	1.058	1.279	0.016
		18	1.472	1.239	1.617	0.007
		24	1.332	1.267	1.427	0.007

It can be seen that the stiffness degrades obviously in the initial stage of cyclic loading, and the performances of all four models are consistent. However, the stiffness degradation index of OMDS still decreases faster with the increase in vibration cycles. Then the degradation model of this paper can better represent this feature of OMDS under cyclic loading. The other three models show that the degradation index tends to stabilize as the vibration cycles increases, which is more suitable for the stiffness degradation characteristic of clay under cyclic loading.

6. Conclusions

The dynamic triaxial test of OMDS under different loads, consolidation ratios, natural moisture contents and wave forms was studied and analyzed, and the conclusions could be drawn as follows:

(1) Under cyclic loading, the soil stiffness decreases with the increase in vibration cycles. The increase in natural moisture content and vibration amplitude, and the reduction in the consolidation ratio accelerates the degradation of soil stiffness. The reason is that a higher moisture content results in the reduction in the interaction force between sand particles so that sand particles are prone to dislocation deformation; a lower consolidation ratio results in the decrease in sand compactness so that the ability to resist elastic deformation becomes weaker.

(2) The degradation index of stiffness decreases rapidly with the increase in vibration cycles, which indicates that the soil can be destroyed with fewer vibration cycles. The degradation index curves of the four load amplitudes are close to each other, which indicates that there is not a critical stress, which is different from cohesive soil. The variation amplitude of the degradation index curve is smaller when the moisture content is lower. The curvature of the degradation index curve with the same consolidation ratio is larger than that with different moisture contents. Among the different wave forms, the rectangular wave has the most obvious effect on the degradation index of OMDS.

(3) A degradation model for OMDS was established. Based on this model, the degradation index under various natural moisture contents and wave forms was fitted and analyzed. The degradation model of this paper can better represent the features of OMDS under cyclic loading.

Author Contributions: Conceptualization, J.D. and Y.Z.; methodology, J.D.; data curation, X.J., B.L. and L.J.; writing—original draft preparation, J.D.; writing—review and editing, J.D. and Y.Z.; funding acquisition, J.D. All authors have read and agreed to the published version of the manuscript.

Funding: This research was funded by the Hainan Province Science and Technology Special Fund, grant number ZDYF2022SHFZ277, the Hainan Provincial Natural Science Foundation of China, grant number 520RC548, the National Natural Science Foundation of China, grant number 52268056, the Hainan Natural Science Foundation, grant number 519QN333, and the Hainan Research Institute of Chinese Academy of Engineering Science and Technology Development Strategy, grant number 21-HN-ZD-02.

Institutional Review Board Statement: Not applicable.

Informed Consent Statement: Not applicable.

Data Availability Statement: Data available on request due to privacy restrictions.

Conflicts of Interest: The authors declare no conflict of interest.

References

1. Du, J.; Liu, B.Y.; Wang, Z.C.; Zheng, G.; Zhou, H.Z. Dynamic Behavior of Cement-Stabilized Organic-Matter-Disseminated Sand under Cyclic Triaxial Condition. *Soil Dyn. Earthq. Eng.* **2021**, *147*, 106777. [CrossRef]
2. Seed, H.B.; Chan, C.K. Clay strength under earthquake loading conditions. *J. Jpn. Soc. Irrig. Drain. Rural Eng.* **1966**, *92*, 53–78. [CrossRef]
3. Wang, J.-Q.; Chang, Z.-C.; Xue, J.-F.; Lin, Z.-N.; Tang, Y. Experimental Investigation on the Behavior of Gravelly Sand Reinforced with Geogrid under Cyclic Loading. *Appl. Sci.* **2021**, *11*, 12152. [CrossRef]
4. Lei, H.Y.; Li, B.; Lu, H.B.; Ren, Q. Dynamic Deformation Behavior and Cyclic Degradation of Ultrasoft Soil under Cyclic Loading. *J. Mater. Civ. Eng.* **2016**, *28*, 04016135. [CrossRef]
5. Cai, Y.Q.; Wu, T.Y.; Guo, L.; Wang, J. Small Stiffness Degradation and Plastic Strain Accumulation of Clay under Cyclic Load with Principal Stress Rotation and Deviatoric Stress Variation. *J. Geotech. Geoenviron. Eng.* **2018**, *144*, 04018021. [CrossRef]
6. He, S.-H.; Zhang, Q.-F.; Ding, Z.; Xia, T.-D.; Gan, X.-L. Experimental and Estimation Studies of Resilient Modulus of Marine Coral Sand under Cyclic Loading. *J. Mar. Sci. Eng.* **2020**, *8*, 287. [CrossRef]
7. Atkinson, J.H. Non-linear soil stiffness in routine design. *Geotechnique* **2000**, *50*, 487–508. [CrossRef]

8. Liang, F.; Zhang, Z.; Wang, C.; Gu, X.; Lin, Y.; Yang, W. Experimental Study on Stiffness Degradation and Liquefaction Characteristics of Marine Sand in the East Nan-Ao Area in Guangdong Province, China. *J. Mar. Sci. Eng.* **2021**, *9*, 638. [CrossRef]
9. Monismith, C.L.; Ogawa, N.; Freeme, C.R. Permanent Deformation Characteristics of Sub-grade Soils due to Repeated Loading. *Transp. Res. Rec.* **1975**, *537*, 1–17.
10. Li, D.; Selig, E.T. Cumulative Plastic Deformation for Fine-Grained Sub-grade Soils. *J. Geotech. Geo-Environ. Eng.* **1996**, *122*, 1006–1013. [CrossRef]
11. Li, D.; Selig, E.T. Method for Railroad Track Foundation Design. II: Applications. *J. Geotech. Geo-Environ. Eng.* **1998**, *124*, 323–329. [CrossRef]
12. Idriss, I.M.; Dobry, R.; Singh, R.M. Nonlinear behavior of soft clays during cyclic loading. *J. Geotech. Eng. ASCE* **1978**, *104*, 1427–1447. [CrossRef]
13. Yao, M.; Nie, S. A model for calculating deformation of saturated soft clay. *J. Hydraul. Eng.* **1994**, *7*, 51–55. (In Chinese)
14. Chai, J.C.; Miura, N. Traffic-load-induced permanent deformation of road on soft subsoil. *J. Geotech. Geoenviron. Eng.* **2002**, *128*, 907–916. [CrossRef]
15. Parr, G.B. *Some Aspects of the Behavior of London Clay under Repeated Loading*; University of Nottingham: Nottingham, UK, 1972.
16. Huurman, M. Development of traffic induced permanent strain in concrete block pavements. *Heron* **1996**, *41*, 29–52.
17. Yasuhara, K.; Hyde, A.F.L.; Toyota, N.; Murakami, S. Cyclic stiffness of plastic silt with an initial drained shear stress. In *Pre-failure Deformation of Geomaterials*; Thomas Telford Ltd.: London, UK, 1998; pp. 373–382.
18. Tan, K.; Vucetic, M. Behavior of medium and low plasticity clays under simple shear conditions. In Proceedings of the 4th International Conference on Soil Dynamics and Earthquake Engineering, Mexico City, Mexico, 23–26 October 1989; pp. 131–141.
19. Wang, J.; Cai, Y.; Xu, C. Experimental study on degradation of stiffness of saturated soft clay under undrained cyclic loading. *Rock Soil Mech.* **2007**, *28*, 2138–2144. (In Chinese)
20. Huang, M.; Li, S. Degradation of stiffness and strength of offshore saturated soft clay under long-term cyclic loading. *Chin. J. Geotech. Eng.* **2010**, *32*, 1491–1498. (In Chinese)
21. Cao, Y.; Kong, L.; Yang, A. Waveform effect of cyclic loading of dynamic character and stiffness degradation characteristics of marine deposited natural soft clay. *Chin. J. Geotech. Eng.* **2013**, *35*, 583–589. (In Chinese)
22. Iraji, A.; Farzaneh, O.; Hosseininia, E.S. A modification to dense sand dynamic simulation capability of Pastor-Zienkiewicz-Chan model. *Acta Geotech.* **2014**, *9*, 343–353. [CrossRef]
23. Hong, Y.; Koo, C.H.; Zhou, C.; Ng, C.W.; Wang, L.Z. Small Strain Path-Dependent Stiffness of Toyoura Sand: Laboratory Measurement and Numerical Implementation. *Int. J. Geomech.* **2017**, *17*, 04016036. [CrossRef]
24. Mei, H.H.; Leng, W.M.; Nie, R.S.; Li, Y.F. Study on dynamic behavior and permanent deformation characteristics of coarse-grained soil. *J. Huazhong Univ. Sci. Technol.* **2019**, *47*, 113–119.
25. Liu, Y.; Zhu, S.Y.; Feng, Y.; Ma, S.K. Research on stiffness softening rules of saturated anisotropic consolidated silty soil under cyclic loading. *Highway* **2020**, *4*, 282–288.
26. Du, J.; Zheng, G.; Liu, B.; Jiang, N.J.; Hu, J. Triaxial behavior of cement-stabilized organic matter–disseminated sand. *Acta Geotech.* **2021**, *16*, 211–220. [CrossRef]
27. Zhang, R.; He, C.; Fei, W.; Gao, M. Effect of consolidation stress ratio on dynamic strength and dynamic pore water pressure of soil. *Chin. J. Geotech. Eng.* **2006**, *28*, 101–105. (In Chinese)
28. Hardin, B.O.; Black, W.L. Closure to vibration modulus of normally consolidated clay: Design equations and curves. *J. Soil Mech. Found. Div. ASCE* **1969**, *95*, 1531–1537. [CrossRef]
29. Wu, Q.; Lu, Q.; Guo, Q.; Zhao, K.; Chen, P.; Chen, G. Experimental Investigation on Small-Strain Stiffness of Marine Silty Sand. *J. Mar. Sci. Eng.* **2020**, *8*, 360. [CrossRef]
30. Liu, J.K.; Cui, Y.H.; Liu, X.; Chang, D. Dynamic characteristics of warm frozen soil under direct shear test-comparison with dynamic triaxial test. *Soil Dyn. Earthq. Eng.* **2020**, *133*, 106114. [CrossRef]
31. Wang, C.; Zhang, J. Experimental and Analytical Study on Residual Stiffness/Strength of CFRP Tendons under Cyclic Loading. *Materials* **2020**, *13*, 5653. [CrossRef] [PubMed]
32. Zhong, H.; Sai, K.V. Normalizing Variation of Stiffness and Shear Strength of Compacted Fine-Grained Soils with Moisture Content. *J. Geotech. Geoenviron. Eng.* **2017**, *143*, 04017058.
33. Huang, J.; Ding, Z.D.; Yuan, T.Y.; Dan, Z.; Li-min, P. Experimental study of dynamic deformation properties of peaty soil under cyclic loading. *Rock Soil Mech.* **2017**, *38*, 2551–2558. (In Chinese)

sustainability

MDPI

Article

Experimental Study on Load-Carrying Behavior of Large Open-Ended Pipe Pile in Cohesionless Soils

Yuan Guo [1,*], Jiale Li [2,*] and Xiong Yu [3]

1 School of Civil Engineering, Sun Yat-sen University, Guangdong 510275, China
2 School of Civil and Transportation Engineering, Hebei University of Technology, Tianjin 300401, China
3 Department of Civil Engineering, Case Western Reserve University, Cleveland, OH 44106, USA
* Correspondence: guoy229@sysu.edu.cn (Y.G.); jiale.li@hebut.edu.cn (J.L.)

Abstract: Large-diameter open-ended pipe piles typically refer to pipe piles with diameters larger than 0.91 m. These piles, when properly designed and constructed, could provide high load-carrying capacity against axial loading, uplift, and overturning in the offshore environment. The current American Association of State Highway and Transportation Officials (AASHTO) specification for pipe piles is based on the database collected from piles with diameters less than 0.61 m and is, therefore, insufficient for the design of large-diameter open-ended pipe piles. This paper introduces an experimental program to investigate the static load-carrying behavior of open-ended pipe piles. Two types of restriction plates, i.e., one-hole plate and four-hole plate, were designed and installed inside the pipe to study their influence on the loading bearing behavior. Beside the laboratory-scale pile experiments, the testing program was augmented with geotechnical centrifuge experiments, which upscale the dimensions of model piles close to the size used at the field scale. The combined laboratory and geotechnical centrifuge experiments help to analyze the load-carrying behavior of pipe piles with different pile diameters, types of restriction plate, and embedment depths. The experimental results indicate that the formation of a soil plug is crucial to mobilize the bearing capacity of pipe piles. The restriction plate helps to form the soil plugging, which leads to a higher end bearing capacity. The selection of types of restriction plate should be determined on a case-by-case basis with considerations of mechanical performance and pile drivability.

Keywords: large-diameter open-ended pipe pile; restriction plate; driven pile; geotechnical centrifuge modeling; pile modeling

1. Introduction

Large-diameter Open-Ended Pipe Piles (LDOEPPs) are steel or prestressed concrete pipe piles with diameters larger than 0.91 m [1]. These piles, when properly constructed, could provide high load-carrying capacity against axial loading, uplift, and overturning, even in poor soil conditions [2,3]. The initial uses and experiences with LDOEPPs are acquired in the offshore industry [4,5]. They are becoming attractive in other areas, such as the transportation engineering, due to savings in construction costs and project schedules [1]. As these piles can be driven into the ground through water, there is no need to build a cofferdam, as is required for the construction of conventional drilled shafts. A US national survey on the use of LDOEPPs conducted by the state Department of Transportations (DOTs) in the US indicated that 44% of state DOTs have had experience with LDOEPPs in recent years [1]. An example application in California showed the use of large open-ended pipe piles for a bridge project had reduced the foundation construction time by half and led to a cost saving of $2.8 million [6].

Despite the advantages in offshore engineering projects, there are also challenges in the design and construction of this type of piling, due to the limited experience at present. As a pile is driven into the ground, soil is squeezed inside the pipe and forms an arching effect within the soil column, which prevents further soil intrusion, but significantly increases the inner unit frictional resistance [7]. The bearing capacity and drivability of open-ended pipe piles depend largely on the degree of soil plugging [8,9]. A fully plugged pipe pile behaves like a closed-ended pile [10]. Experience also indicates that this type of pile cannot be treated merely as a scaled-up version of a standard driven pile. Bradshaw et al. [11] measured the load transfer curves from a static load test performed on a 1.8 m-diameter pipe pile in silty soils in Rhode Island and reported a smaller loading carrying capacity compared to conventional piles.

Various experimental studies have been conducted aiming to understand the soil plug formation mechanism inside LDOEPPs [12–14]. By using photo-elastic materials, Dijkstra and Broere [15] demonstrated the force transmission and arching effect at the bottom of a soil plug. Laboratory experiments also indicated that there is a significantly high load transfer over the bottom of a soil plug, which is about three to four times the pile diameter [16,17]. Field tests on large-diameter pipe piles have shown that a relatively large displacement is required to fully mobilize the plug resistance [11]. Numerical methods, such as the finite element method and discrete element method, were also used to study the mechanical performance of open-ended pipe piles under axial loads [18–21]. Factors determining the plugging mechanism include: hammer-pile impact, construction disturbance (driven pile or jacked pile), pile diameter, pile length to diameter ratio, mechanical soil properties, and friction at the soil–pile interface [22–26]. There are also studies focusing on the potential corrosion of steel materials in long-term service life [27].

Recently, engineers and researchers have proposed various geometric designs, including the use of pile sleeves, for LDOEPPs to control the formation of soil plugging inside the pipe pile. Kumara et al. [28] showed that the installation of a small inner sleeve at the bottom of pipe piles could increase the degree of soil plugging. Kien and Ito [29] implemented an external spiral wing (700 mm in diameter) at the bottom of a 355 mm-diameter steel pipe pile to improve the bearing capacity. In this study, the soil plugging mechanism of pipe piles in cohesionless soils, particularly in relation to the pipe diameter, is presented. Two innovatively designed restriction plates, i.e., a one-hole plate (Figure 1) and a four-hole plate, were installed inside the model pipes and compared by conducting laboratory and geotechnical centrifuge load tests. The results may help to further the understanding of the load carry behavior of LDOEPPs.

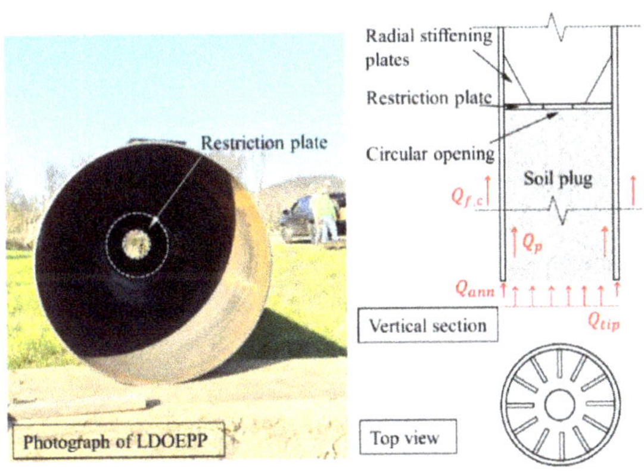

Figure 1. Photograph (**left**) and schematic (**right**) of a large-diameter open-ended pipe pile with a restriction plate designed by the authors.

2. Current Design Practice

This section summarizes seven commonly used design methods for LDOEPPs to facilitate a brief understanding of the current design practice. In general, the soil plugging effect determines the load-carrying behavior of LDOEPPs and, therefore, should be properly considered in the pile design.

2.1. Soil Plugging Effect in Pipe Pile Design

Most of the design experience of LDOEPPs are accumulated in the offshore industry [2,4,30]. The American Petroleum Institute (API) recommends that the total axial resistance is the sum of the external shaft friction and the end bearing on the pipe wall annulus plus the total internal shaft friction (i.e., plug resistance) or the end bearing of the pile, whichever is the less, as given in Equation (1) [31]:

$$Q_c = Q_{f,c} + \min\{Q_{tip}, Q_{ann} + Q_p\} \quad (1)$$

where Q_c is the total resistance of an open-ended pipe pile; $Q_{f,c}$ is the shaft friction provided by the outer soil–pile interface; Q_{tip} is the tip resistance of the whole cross section; Q_{ann} is the resistance of the pipe tip annulus; and Q_p is the plug resistance.

As the open-ended pipe pile is driven underground, soils are squeezed into the pipe and form a soil column inside, called the soil plug. There are frictional contact forces on the pile–soil plug surface, which mobilize soil particles moving downwards. As these soil particles inside the soil plug are confined by the pipe wall, the pile driving process squeezes the soil column to form a densely-packed soil plug, which significantly increases the normal contact force on the pile–soil plug surface, and, in turn, leads to higher frictional forces. Under certain conditions, the soil column inside the pipe pile may be fully plugged and the driving and bearing behavior of an open-ended pipe pile is like that of a close-ended pipe pile.

In the current design practice, the plugging effect is usually considered in terms of two factors, pile dimension and soil mechanical strength [32–35]. Some additional parameters that can be acquired during the pile installation, such as the Incremental Filling Ratio (IFR) and the Plug Length Ratio (PLR), have been incorporated to describe the degree of soil plugging [14,36]. The typical definitions of IFR and PLR for LDOEPPs are provided in Equation (2). When the IFR equals zero as the length of soil plug remains constant under driving, the pipe pile is considered to be fully plugged and the base resistance is regarded as equivalent to a closed-ended pile. The API design practice also indicates that the plugged or unplugged condition needs to be verified, based on static analyses, i.e., a pile can be driven in an unplugged condition but behaves as plugged under static loads [31].

$$IFR = \Delta L_p / \Delta L, \ PLR = L_p / L \quad (2)$$

where ΔL_p is the plug length increment; ΔL is the pile driving increment; L_p is the length of soil plug; and L is the driving length of pipe pile.

2.2. Comparison of Different Design Methods

A literature review was conducted on the state of practice of the design methods for LDOEPPs [37]. Table 1 compares seven major design criteria of LDOEPPs for offshore and transportation projects, which include the API design practice [31], the Federal Highway Administration (FHWA) recommended method [38], the Norwegian Geotechnical Institute (NGI) method [39], the Finnish National Road Administration (FinnRA) method [40], the China Ministry of Construction (CMC) code for deep foundations [41], the Imperial College Pile (ICP) method [42], and the University of Western Australia (UWA) method [43]. These design criteria can be classified into two groups depending on the different descriptions of soil plugging: (1) the explicit approach, that considers annulus tip resistance as well as plug resistance, such as the API design practice, the FHWA method, and the NGI method;

and (2) the implicit method, that considers equivalent tip resistance, such as the FinnRA method, the CMC Code, the ICP method, and the UWA method.

For the API, FHWA, and NGI methods, the plug resistance and annulus resistance are calculated and compared with the tip resistance within the total cross-sectional area of the pile. The soil plugging state is determined and the smaller value is used as the pile tip resistance. However, there are major differences in determining the plug resistance, as the formation mechanism of the soil plug is still unclear. The inner unit shaft friction in the NGI method is assumed to be three times larger than the outer unit friction considering soil arching. The API method assumes the inner and outer unit shaft frictions are the same, while in the FHWA method, the inner unit friction is assumed to be 0.33 to 0.50 of the outer unit friction, and the annulus unit resistance is also reduced by half [31,38,39].

Table 1. Comparison of different design methods for LDOEPPs.

Method	Pile Tip Resistance	Basic Principle	Feature	Determinant Factors
API Design Practice [31]	Lesser of annulus + plug resistance and tip resistance	Soil strength acquired through laboratory or in-situ test	Inner unit shaft friction equals outer unit friction	Soil property
FHWA Method [38]			Inner unit shaft friction is 1/2~1/3 of outer unit friction; reduction in unit tip resistance	Soil property, pile diameter, pile shoe configuration
NGI Method [39]		CPT based approach	Inner unit shaft friction is three times outer unit friction in sand	Cone tip resistance, relative density, etc.
FinnRA Method [40]	Equivalent tip resistance	A plugging coefficient to describe degree of soil plugging	Fully plugged as pile length-diameter ratio (L/D) > 10 in moraine, and L/D > 15 in sand	Soil property, pile length-diameter ratio
China Deep Foundation Code [41]			Fully plugged as L/D > 5	Pile length-diameter ratio
ICP Method [42]		CPT based approach	CPT data are used to classify plugged or unplugged pile	Cone tip resistance, pile diameter, relative density, driving depth
UWA Method [43]			IFR is used to describe plugging mode during installation	Cone tip resistance, pile diameter, IFR

In the FinnRA and CMC methods, the soil plugging effect is considered implicitly by implementing a plugging coefficient for the pile tip resistance, as shown in Table 1. The plugging coefficient is determined by the pile length–diameter ratio (L/D), and a fully plugged pile has a plugging coefficient of 0.8 [40,41]. From the CMC code, pipe piles with a length–diameter ratio larger than 5 is assumed to be fully plugged, while in the FinnRA method, the critical pile length–diameter ratio is 10 in moraine conditions and increases to 15 in sand. These two methods are easy to implement but lack the consideration of complex soil conditions, e.g., various soil types.

Both the ICP and UWA methods are based on the in-situ Cone Penetration Test (CPT) results [42,43]. In the ICP method, piles are divided into plugged or unplugged mode according to the pile diameter and soil condition. The tip resistance is assumed to be proportional to the cone tip resistance as measured in the CPT test. In the UWA method, the IFR has been used as a way of correction for tip resistance. However, in-situ large scale tests provide evidence that the CPT-based method provides a reliable prediction on the skin friction while the accurate estimation of the end bearing capacity is still challenging [44].

3. Experimental Program

The load-carrying behavior of pipe piles in cohesionless soils was investigated by conducting laboratory-scale and geotechnical centrifuge experiments. The experimental program, that combined different pile configurations, i.e., different pile diameters, types of restriction plates, and embedment depths, is introduced in this section.

3.1. Laboratory-Scale Pile Static Load Test

As the field-scale pile load tests are time-consuming and inefficient, we started the tests with small diameter pipe piles in the laboratory to study the bearing behavior of pipe piles with the customized restriction plate. There is no scaling-up implemented in this part of work. For the laboratory-scale experiment, a series of tests were conducted on pipe piles of three different diameters (i.e., 51-mm, 102-mm, and 152-mm in diameter) in combination with different configurations of the restriction plate (i.e., no restriction plate, restriction plate with one circular hole, restriction plate with four circular holes), as shown in Figure 2. The restriction plates were designed to provide additional confinement to the soil plug to increase the inner frictional force. The area of one of the holes was equivalent to the area of four holes.

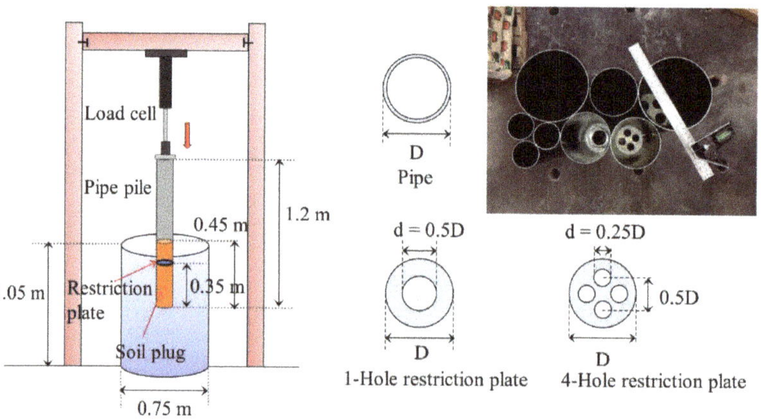

Figure 2. Setup of laboratory-scale experiment of pipe piles with different restriction plates.

The geometric dimensions of pipe piles tested in the laboratory-scale experiment are listed in Table 2. The tests were conducted with one case for each scenario. The restriction plates were cut using a water jet cutter and glued to the specified location with epoxy at a height of 0.35 m from the bottom. A position of 0.35 m for the restriction plate corresponded to a further driving depth of 0.10 m after the soil plug reached the restriction plate surface. It was shown that this was enough to achieve a fully mobilized soil plug without disturbing the stress concentration at the pile tip. A customized loading system was developed for the laboratory pipe pile test to provide enough vertical force without loading eccentricity. The maximum vertical loading capacity of the load cell was 17.8 kN and the maximum displacement range of the loading piston was 17.8 cm, meeting the experimental needs in our trial test.

The cylindrical sand box, that had a diameter of 0.75 m and a height of 1.05 m, is shown in Figure 2. The air-pluviation method was used in preparing the soil samples, in which sand grains were poured layer by layer through a single pipe with a constant pouring height of around 30 cm and a steady deposing intensity. After pouring sand into the sand box, the soil sample was slightly densified by tapping the box walls uniformly with a rubber hammer to achieve a relative density of 74% and a density of 1.604 g/cm^3, which were the same as used in the centrifuge tests. After each loading test, the soil samples were removed and refilled with the same preparing procedure. Before each test, the pipe piles were driven into the sand manually by dropping a 10 kg hammer from a height of 15 cm to simulate the pile driving, and the final embedment depth was 0.45 m with a variance less than 1 mm. The thickness of the bottom sand layer below the pile tip was at least four times that of the pile diameter, and the diameter of the drum was at least five times that of the pile diameter, which helped to minimize the potential boundary effect [45]. It should be noted that the pile installation would densify the sands around the pile, and its influence has been neglected for simplicity in this study.

Table 2. Geometric dimensions of model piles for laboratory-scale experiment (restriction plate is located 35 cm above pile tip, and embedment depth is 45 cm for all piles).

Pile No.	Plate Type	Inner Diameter, mm	Wall Thickness, mm	Diameter of Plate Holes, mm
1	Pipe	50.78	1.55	N.A.
2	One-hole	50.78	1.55	25.39
3	Four-hole	50.78	1.55	12.70
4	Pipe	99.09	1.49	N.A.
5	One-hole	99.09	1.49	49.55
6	Four-hole	99.09	1.49	24.77
7	Pipe	148.76	2.01	N.A.
8	One-hole	148.76	2.01	74.38
9	Four-hole	148.76	2.01	37.19

A stepwise loading sequence was applied on top of the pile to mobilize the ultimate resistance of pipe piles [46]. The increment of loading levels was calculated by estimating the failure resistance (at movement exceeds 15% of pile diameter) divided by fifteen loading levels. Once prepared, the vertical loading force was applied to the pipe pile to drive it into the sand box with a constant velocity of 1.0 mm/s until reaching the preset force value. The loading frame had a maximum limit of 25,000 N for the axial loading force with considerations of possible buckling. From trial tests, the model piles achieved the full displacement in dry sand quickly, in less than two minutes (around 100 mm cap displacement). To ensure reaching the steady state, each loading level was maintained for 5 min after reaching the preset force value.

For LDOEPPs, a relatively large cap displacement is required to fully mobilize the pile's base resistance. The conventional Davisson offset method may underestimate the base resistance, especially for piles with diameters larger than 0.6 m. For this reason, some agencies (e.g., Florida DOT) use a modified Davisson offset method to determine the ultimate bearing capacity of piles that are 0.6 m or larger [1]. The displacement at the strength limit is defined as the offset of $D/30$ plus the elastic deformation of piles, where D is the pile diameter, see Equation (3). This modified Davisson's method was used to calculate the bearing capacity of the pipe piles. However, it should be noted that agencies may choose more conservative methods in the pipe pile design.

$$S_f = PL/AE + D/30 \tag{3}$$

where S_f is the maximum allowable pile displacement, in m; P is the axial load, in N; L is the pile length, in m; A is the cross-section area of the pile, in m^2; E is the elastic modulus of the steel pipe, in Pa, using a value of 200 GPa in this study; and D is the pile diameter, in m.

3.2. Centrifuge Pile Static Load Test

The geotechnical centrifuge experiment was conducted to scale up the pile dimensions to approximate field sizes. The centrifuge experiment tests small-scale models under high gravitational levels to simulate the stress state of field conditions [47]. By using the centrifuge, the dimensions of the prototype models (e.g., pile diameter and pile length) were scaled up by the applied gravity acceleration, according to established scaling laws, as shown in Table 3 [48]. For example, a 1 cm-diameter model pile tested under a gravity of 100 g represents the behavior of a 1 m-diameter prototype pile under gravity of 1 g.

Table 3. Scale factors for geotechnical centrifuge modeling.

Quantity	Symbol	Units	Scale Factor
Length	L	L	1/N
Volume	V	L^3	$1/N^3$
Mass	M	M	$1/N^3$
Gravity	g	L/T^2	N
Time	T	T	1/N
Force	F	ML/T^2	$1/N^2$
Moduli	E	M/LT^2	1
Strength	s	M/LT^2	1

The geotechnical centrifuge at Case Western Reserve University (CWRU) was used for this study. The centrifuge payload capacity was a 20 g-ton with a maximum acceleration of 200 g for static tests and 100 g for dynamic tests. The internal dimensions of the specimen box were 53.3 cm (length) by 24.1 cm (width) by 17.7 cm (height). Similar to the laboratory experiments, pipe piles with different diameters, in combination with different types of restriction plates, were tested for their bearing performance. Pipe pile models were made of steel and the restriction plates were made from 3D printed polylactic acid to achieve the most accurate dimensions. The design of the restriction plates were the same as that shown in Figure 2.

The model setup of the geotechnical centrifuge experiment is shown in Figure 3. The customized loading system consisted of an actuator, a load cell, and a linear variable differential transformer. The maximum range of the load cell was 1330 N. Nine model pipe piles were tested under different centrifugal accelerations of 1 g, 10 g, 30 g, and 50 g, respectively. The 1 g test was conducted in the soil container without spinning. The geometric dimensions of the model piles are provided in Table 4. The inner diameters of these piles were 19.0 mm, 25.4 mm, and 38.4 mm. The embedment depth of the model pile was 8 cm. As the model piles were scaled up by different factors, the embedment depth of prototypes would be different. The wall thickness influenced the soil squeezing inside the pipe. In this study, we assumed the influence of wall thickness on soil plugging was negligible as it is generally much smaller compared to the pile inner diameter. For each of the pile diameters, a one-hole or four-hole restriction plate was included for comparison with the open-ended pipe pile. The restriction plates were installed 4.0 cm up from the pile tip. Due to the limitation of the soil container in height, the embedded depth of the pile was set at 8.0 cm in this study so that the restriction plate was engaged with soil with a significant driving distance of 4.0 cm before the load test.

The air-pluviation method was used in soil sample preparation. The pipe pile model was installed by hammer driving to simulate the installation in field conditions. The installation was conducted under a 1 g environment. Before the loading process, the specimen box was spinning to the target acceleration to reach the balance under gravity. Similar to the 1 g laboratory-scale test, a stepwise load was applied on top of the model piles under target centrifugal accelerations. There are studies showing the installation method may lead to different pile bearing behaviors in the centrifuge modeling [49]. Li et al. [50] compared the traditional pile installation at 1 g with the in-flight installation at 50 g, and concluded that the in-flight pile installation leads to a denser neighboring soil and a higher bearing capacity. In this paper, the potential influence of the pile installation method and the changing soil relative density were not considered. To achieve the ultimate bearing capacity in a short time period, there were six loading stages and four unloading stages implemented during the centrifuge test. The magnitude of each loading stage was determined by a trial test. The loading force was applied through the actuator with a self-defined servo mechanism, and usually it took less than 15 s to achieve the target force.

The cap displacement and load force were recorded in real time during the test, and the steady state was determined case-by-case based on the cap displacement. Each loading state was held for around 45 s to ensure the pile reached the steady state.

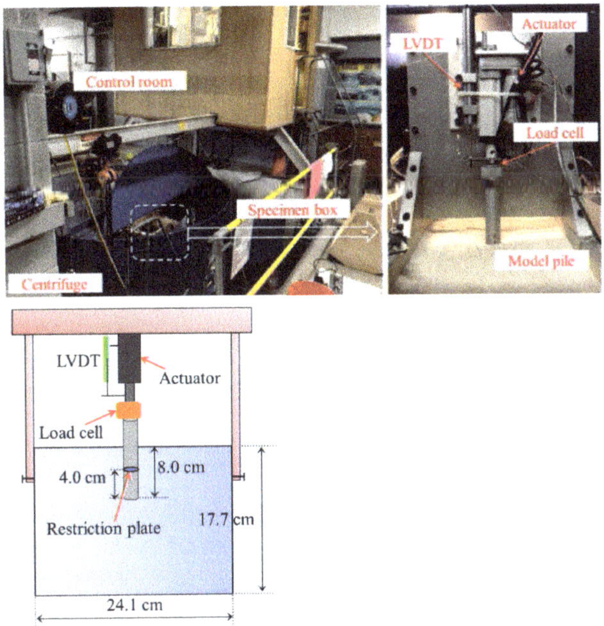

Figure 3. CWRU 20 ton-g geotechnical centrifuge machine and test setup (LVDT: linear variable differential transformer).

Table 4. Geometric dimensions of model piles for centrifuge experiment (restriction plate is located 4 cm above pile tip, and embedment depth is 8 cm for all piles; all the piles were tested under 1 g, 10 g, 30 g, and 50 g).

Pile No.	Plate Type	Inner Diameter, mm	Wall Thickness, mm	Diameter of Plate Holes, mm
10	Pipe	19.0	0.9	N.A.
11	One-hole	19.0	0.9	9.50
12	Four-hole	19.0	0.9	4.75
13	Pipe	25.4	0.9	N.A.
14	One-hole	25.4	0.9	12.70
15	Four-hole	25.4	0.9	6.35
16	Pipe	38.4	0.9	N.A.
17	One-hole	38.4	0.9	19.20
18	Four-hole	38.4	0.9	9.60

3.3. Soil Properties

The soil used in the testing program was silica sand provided by Oglebay Norton Industrial Sand Inc., a commercial supplier in north Ohio. The maximum and minimum bulk densities were 1.67 g/cm^3 and 1.43 g/cm^3, respectively. The mean particle diameter was 0.20 mm from the sieve analysis, with a uniformity coefficient of 2.10 and a coefficient of gradation of 1.22. The sand was classified as poorly graded sand (SP), based on the unified soil classification system [51]. The ratio of pile inner diameter on the sand's D_{90} was 68 for the 19.0 mm-diameter model pile. A direct shear test was conducted to measure the internal friction angle of sand specimens under dry conditions, and the normal stress ranged from 50 kPa to 200 kPa. With a relative density of 75% prepared in the direct shear test, which was similar to the sand sample used for the pile load test, the deduced internal friction angle was 31.6°.

4. Experimental Results

The testing results of the laboratory and centrifuge pile loading experiments are presented in this section. The ultimate bearing resistance of pipe piles with different pile diameters and types of restriction plates was compared.

4.1. Ultimate Resistance from Laboratory-Scale Experiment

The length of soil plug was recorded before and after the pile load test by measuring the surface height of soils inside the pipe pile, and the results are shown in Figure 4. In general, a shorter soil plug, as in Figure 4a, indicated a more compacted soil column after pile installation. If there was no plugging, the soil plug length should equal the penetration depth, i.e., 45 cm. If the pipe pile was fully plugged, the soil plug length should be much lower than 45 cm, as was the case for all the 51-mm piles and 102-mm piles with restriction plates. During the pile installation, it was observed that the restriction plate was effective in facilitating the soil plugging, especially for 51-mm and 102-mm piles. For the 152-mm piles, the length of soil plug was similar with or without the restriction plate. Figure 4b plots the measured soil plug length after the load test. As the piles were driven downward, the soil plug length was expected to increase, except for the fully plugged condition, in which it would remain the same. By comparing the plug length increment before and after the pile load test, one can see that all the 51-mm piles were in a fully plugged condition, while only the 102-mm piles with restriction plates were fully plugged, which was consistent with our discussion from Figure 4a. For the 152-mm piles, the soil plug length increased for pipe pile and the pile with 1-hole plate, indicating an unplugged condition.

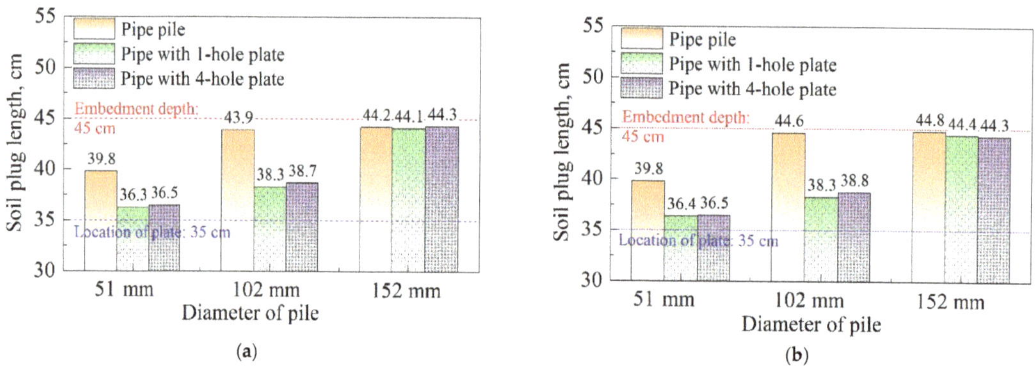

Figure 4. Measured soil plug length from experiments: (**a**) before load test; and (**b**) after load test.

The load–displacement curves of different-sized piles are provided in Figure 5. The modified Davisson's offset [1] was plotted as well, based on Equation (3). The intersect between the load–displacement curve and the offset line determines the ultimate bearing capacity of pipe piles. As the elastic deformation of model piles was small, the Davisson's offset approximates a horizontal line in Figure 5. Each loading curve represented one testing case under that scenario. For the 51-mm piles shown in Figure 5a, all these piles were fully plugged, and the bearing capacities were similar. The effectiveness of restriction plate was more obvious for the 102-mm and 152-mm piles. With the installation of restriction plates, the ultimate bearing capacities of 102-mm piles increased from 2.9 kN to 4.2 kN (or by 45%) with the one-hole restriction plate, and to 6.5 kN (or by 124%) with the four-hole restriction plate. Similar observations existed for the 152-mm piles, as shown in Figure 5c. With the installation of the restriction plate, the soil plug was confined vertically, leading to an increase in inner frictional force on the pile-soil plug surface. The four-hole restriction plate provided better confinement than the one-hole restriction plate. In general, the use of a restriction plate significantly improved the load bearing capacity, and the four-hole restriction plate was more effective than the one-hole restriction plate with the same opening plate area.

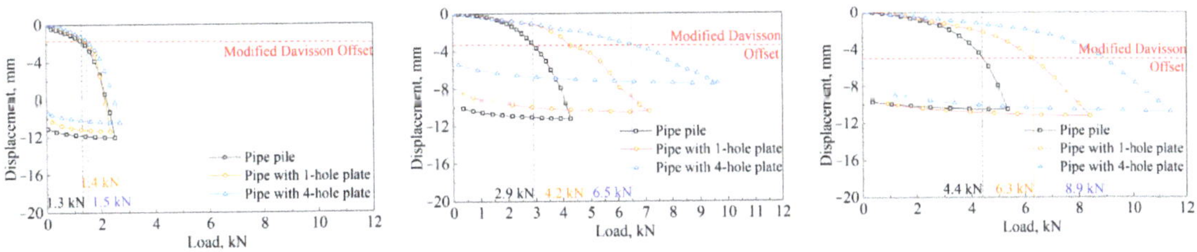

(a) (b) (c)

Figure 5. Load–displacement curves of pipe piles in laboratory load tests: (**a**) 51-mm diameter piles; (**b**) 102-mm diameter piles; and (**c**) 152-mm diameter piles.

4.2. *Ultimate Resistance from Scaled Centrifuge Experiment*

The soil plug length was been recorded in the centrifuge experiment, as the manual installation of small model piles might introduce variances, which could be large by scaling up the pile dimensions. Figure 6 plots three examples of load–displacement curves of different-sized model piles under 50 g. The axial load was scaled by a factor of 2500, while the cap displacement was scaled by a factor of 50 following Table 3. The modified Davisson's offset was also plotted in these figures, following Equation (3), in which it consists of an offset based on pile diameter and an offset for elastic deformation. The inclination was relatively small due to the small elastic deformations of these piles. Compared to the laboratory-scale experiments, there were similar observations in terms of the ultimate bearing capacity of the pipe piles. The use of a restriction plate confined the vertical motion of the soil plug which led to a higher ultimate bearing capacity, and the piles with a four-hole restriction plate had the best load carrying performance.

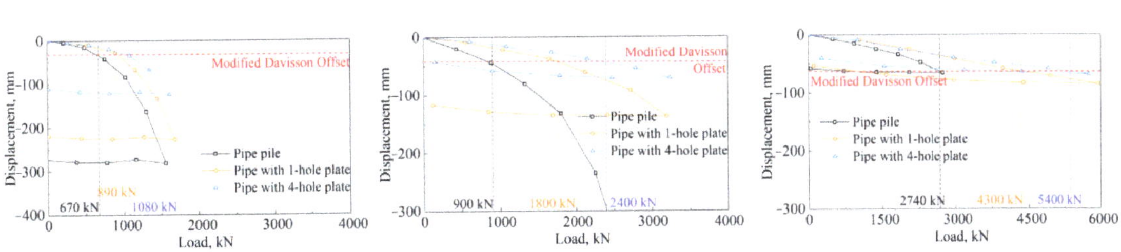

(a) (b) (c)

Figure 6. Examples of load–displacement curves of pipe piles in centrifuge experiments under 50 g: (**a**) 19.0 mm-diameter model piles (0.95 m-diameter prototype); (**b**) 25.4 mm-diameter model piles (1.27 m-diameter prototype); and (**c**) 38.4 mm-diameter model piles (1.92 m-diameter prototype).

The ultimate bearing capacities of prototype piles (after applying the scaling law) under different centrifugal accelerations (i.e., 1 g, 10 g, 30 g, and 50 g) are listed in Appendix A. The laboratory-scale and centrifuge experiments contain data that represent the load-carrying behaviors of pipe piles of different pile diameters, types of restriction plates, and embedment depths. The location of restriction plate was also different for different prototype piles. In the design practice, the location of restriction plate was mainly determined by the soil conditions, e.g., in northern Ohio, the restriction plate was placed one meter below the bearing dense sand layer. In these experiments, the location of the restriction plate was designed below the embedment depth with a sufficiently long mobilizing distance during the pile driving. As the soil plug was fully mobilized after driven and soil plug confinement was more obvious at the bottom pile, well below the restriction plate, the degree of soil plugging was similar for piles with different restriction plate locations. Considering this, these experimental data were compiled in Appendix A without considering the difference in restriction plate locations.

It was also noted that the measured ultimate bearing capacity was slightly lower than the data reported by other researchers. The primary reason was the different relative density of the sand samples. The sand samples prepared through the air-pluviation method tended to form a loose packing state, while the ultimate bearing capacity of the pipe pile was quite sensitive to the relative density. Besides, the embedment depth also influenced the skin friction and the tip resistance.

5. Further Discussions

In this section, the experimental data from the laboratory-scale and centrifuge experiments are synthetized and analyzed by comparing the normalized bearing capacity and equivalent tip resistance over different pile diameters. A soil plug coefficient, which describes the degree of soil plugging inside the pipe pile, is also proposed.

5.1. Normalized Bearing Capacity

To compare the bearing capacity of pipe piles of different diameters, a simplified normalization process was used by dividing the bearing capacity by the cross-section area (to determine unit bearing capacity) and the vertical earth pressure at the pile tip (to account for confinement effects by the embedment depth), as given in Equation (4). Figure 7 plots the normalized unit area bearing capacity versus the pile diameter. A fully plugged open-ended pile behaved as a close-ended pile with a higher end bearing capacity. As a result, the normalized unit area bearing capacity was much higher for small diameter piles. As the pile diameter increased, the bearing capacity was provided by the inner frictional force and the normalized unit area bearing capacity became stable. The figure also shows that large-diameter pipe piles with a restriction plate generally had a high normalized unit area bearing capacity, particularly for pipe piles with four-hole restriction plates.

$$q_{u,norm} = Q_u/(A\rho gH) \quad (4)$$

where $q_{u,norm}$ is the normalized unit area ultimate bearing capacity of the pile; Q_u is the ultimate bearing capacity derived from the modified Davisson's criteria; A is the cross-section area of the pile; ρ is soil bulk density; g is the gravitational acceleration; and H is the embedment depth of the pile.

The variance of normalized bearing capacity over pile diameter is primarily due to the different bearing mechanisms of pipe piles. For a small diameter pipe pile, it is easier to form a fully plugged condition after driven. As it behaves as a close-ended pile, the normalized bearing capacity is generally higher. As the diameter increases, the bearing capacity is provided by the frictional force on the inner pile surface, which is smaller than the direct bearing resistance of soils at the pile toe. As a result, the normalized bearing capacity decreases over pile diameter.

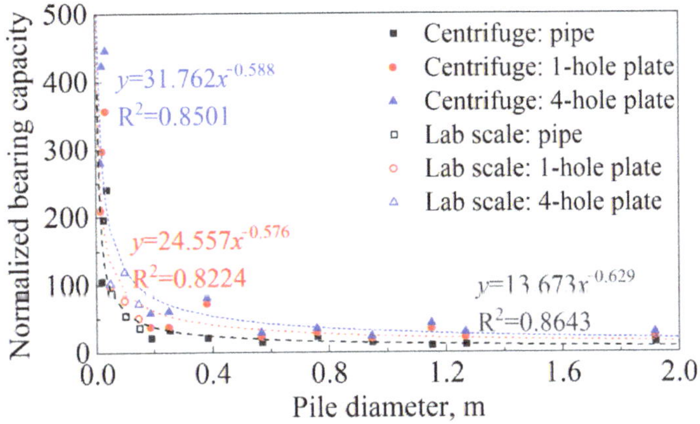

Figure 7. Normalized bearing capacity of prototype piles over pile diameter.

5.2. Equivalent Tip Resistance

The above normalization does not separate shaft resistance from the ultimate bearing capacity. Therefore, the concept of equivalent tip resistance in the FinnRA [40] and CMC [41] methods was adopted to better quantify the load-carrying performance of pipe piles over different pile diameters. The equivalent tip resistance was calculated by subtracting the external shaft resistance from the total bearing capacity. As the ratio of pile embedment depth to pile diameter is smaller than 10.0, a triangular earth pressure distribution is assumed to estimate the external shaft resistance, as given in Equation (5) [52]. The frictional angle on soil-steel interface was 18.1° as measured from the direct shear test, in which the bottom shearing box was replaced with a steel plate with similar surface roughness as the pipe pile, and the top shearing chamber was filled with sand.

$$f_s = tan\delta \cdot k_s \rho g H \tag{5}$$

where f_s is the unit frictional resistance of the external shaft; δ is the frictional angle of soil-pile interface; and k_s is the coefficient of lateral earth pressure.

The sorted data on equivalent tip resistance over pile diameter are visualized in Figure 8. With the installation of a restriction plate, a remarkable improvement in end bearing capacity was obtained. The average increment in equivalent tip resistance for pipe piles with one-hole restriction plates was 91% compared to the same-sized pipes, and increases to 155% for pipe piles with four-hole restriction plates, as also see Appendix A. Figure 9 plots the equivalent tip resistance of pipe piles versus the pile diameter. The pile piles with diameters smaller than 0.15 m were not incorporated as they are not commonly used in engineering practice. The Terzaghi's theoretical bearing capacity was also plotted for comparison, see Equation (6) for details:

$$q_u = 0.5\rho g D S_\gamma N_\gamma + \rho g H N_q \tag{6}$$

where q_u is the theoretical ultimate bearing resistance; D is the pile diameter; N_γ and N_q are Terzaghi's bearing capacity factors, and were 20.7 and 33.0 from Meyerhof's recommendation; S_γ is the shape correction factor; H is the embedment depth of model piles.

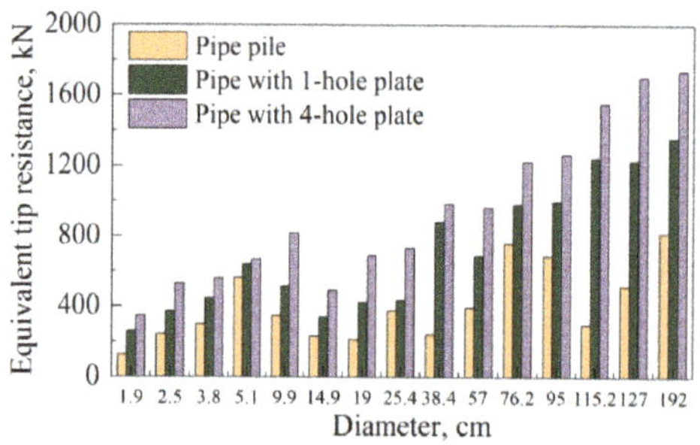

Figure 8. Histogram of equivalent end bearing capacity of different prototype piles.

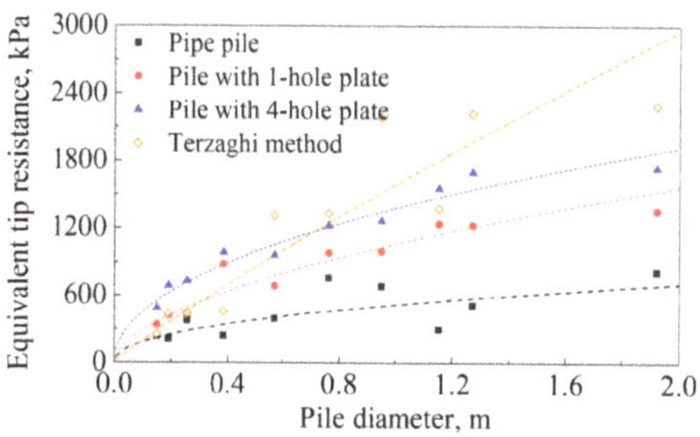

Figure 9. Equivalent end bearing capacity of prototype piles over different pile diameters.

For pipe piles of a given diameter, Terzaghi's bearing capacity theory estimates a linear relationship between the theoretical bearing resistance and the embedment depth (see Equation (6)). Figure 10 compares the equivalent tip resistance of different model piles over different embedment depths. For piles with a small embedment depth, the contribution on tip resistance mainly came from the first term in Equation (6), which is determined by pile diameter. As the embedment depth increased, the second term (determined by embedment depth) dominated the tip resistance, and the variance of equivalent tip resistance for large embedment depth piles, in Figure 10, became small. It was also observed that the equivalent tip resistance did not follow a linear relationship with the diameter or the embedment depth, as the resistance of the LDOEPPs was provided by the inner pipe friction rather than the tip bearing assumed by Terzaghi's theory.

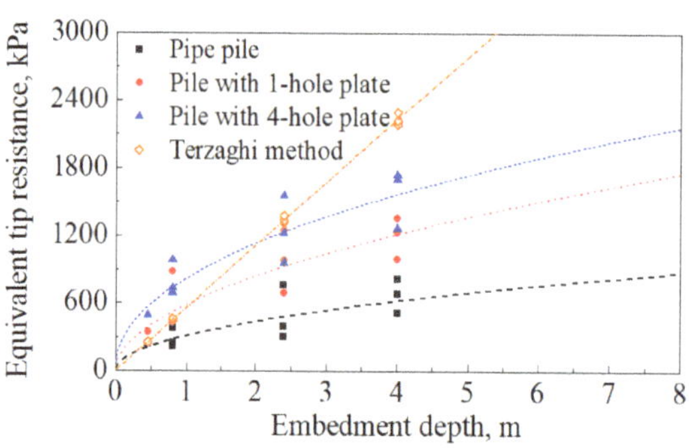

Figure 10. Equivalent end bearing capacity of prototype piles over different embedment depths.

5.3. Soil Plug Coefficient

The end bearing capacity of open-ended pipe piles depends highly on the degree of soil plugging. A fully plugged pipe pile has similar behavior to a close-ended pipe pile, and the bearing capacity can be estimated by Terzaghi's theoretical bearing resistance. As the pile diameter increases, the equivalent end bearing capacity decreases and there are relative motions between the pile and soil plug. The degree of soil plugging can be described by a defined soil plug coefficient, which equals the equivalent tip resistance divided by the theoretical ultimate bearing resistance, as is given in Equation (7). Values of the soil plug coefficient range from zero to one, and a fully plugged pipe pile corresponds to a value of one. The calculated soil plug coefficient of the model piles is listed in Appendix A. For LDOEPPs driven in dense sands, the pile diameter is one of the most important determining factors on the pile's plugging state. Figure 11 plots the calculated soil plug coefficients over pile diameters ranging from 0.15 m to 1.92 m. It is shown that the soil plug coefficient for small-diameter piles might exceed 1.0, indicating a much higher end bearing resistance than the theoretical calculation. A possible explanation was that the fully plugged pipe pile behaved as a closed-ended pile. As the pile was driven downwards for a significant distance, soils around the pile tip were densified, which introduced additional resistance on the pile tip.

$$\alpha = q'_{u,tip}/q_u \tag{7}$$

where α is the soil plug coefficient; $q'_{u,tip}$ is the equivalent tip resistance; and q_u is the theoretical ultimate bearing resistance.

The critical diameter for pipe piles is defined as the pile diameter corresponding to a soil plug coefficient of one. Below the critical pile diameter, the pipe pile is regarded as fully plugged. The critical pile diameter of pipe piles without restriction plates was 0.08 m, and increased to 0.30 m for pipe piles with one-hole restriction plates and 0.70 m for pipe piles with four-hole restriction plates. Equations (8)–(10) give the power law regressions between soil plug coefficient and pipe pile diameter, taking into consideration the critical pile diameter. With these equations, the equivalent tip resistance of LDOEPPs with different restriction plates driven in sand could be estimated, and the total bearing capacity of the pile equaled the equivalent tip resistance plus the shaft resistance. It should be noted that the regressed soil plug coefficient showed a high variance due to limitations in experimental setups, and each test was conducted once without a parallel case. The primary goal of the soil plug coefficient model is to provide a simplified approach for bearing capacity estimation; however, due to limitations of this study, the usage should be determined on a case-by-case basis.

For open-ended pipe piles:

$$\begin{gathered} \alpha = 0.3276D^{-0.456},\ D > 0.08\ \text{m} \\ \alpha = 1.0,\ D < 0.08\ \text{m} \end{gathered} \tag{8}$$

For open-ended pipe piles with one-hole restriction plates:

$$\begin{gathered} \alpha = 0.6693D^{-0.342},\ D > 0.30\ \text{m} \\ \alpha = 1.0,\ D < 0.30\ \text{m} \end{gathered} \tag{9}$$

For open-ended pipe piles with four-hole restriction plates:

$$\begin{gathered} \alpha = 0.8688D^{-0.424},\ D > 0.70\ \text{m} \\ \alpha = 1.0,\ D < 0.70\ \text{m} \end{gathered} \tag{10}$$

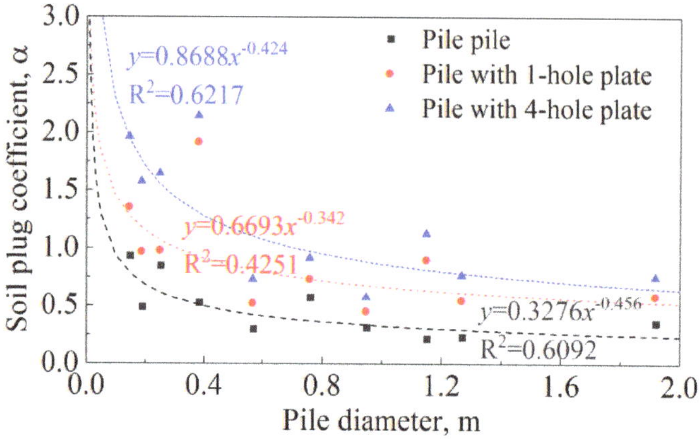

Figure 11. Calculated soil plug coefficient of prototype piles over pile diameters with and without a restriction plate.

5.4. Limitations of Current Research

While this study covered a wide range of pile configurations in cohesionless soils with/without restriction plate, there were a few simplifications adopted, considering the complexity of this problem. In the geotechnical centrifuge experiment, the potential influence of particle size and relative density was generally ignored in scaling up the model piles. With the implementation of restriction plates, soil particles are confined in the vertical motion, which means soil arching forms easily, as is shown in Figure 12. Current experience indicates that the arching effect is also determined by soil grain size and relative density. Further study is needed to improve the current understanding. A comparison between the geotechnical centrifuge test and full-scale experiment may also be helpful in clarifying this. Due to equipment limitations, the embedment depths of prototype piles were within 5.0 m. Here we assumed the bearing mechanism of restriction plates was similar over different embedment depths as the soil plug was fully mobilized. The proposed soil plug coefficient simplified the estimation of pile end bearing capacity. However, due to the complexity of soil plug formation and the limitations of this study, there is still room for improvement, as can be seen from the high variance shown in Figure 11.

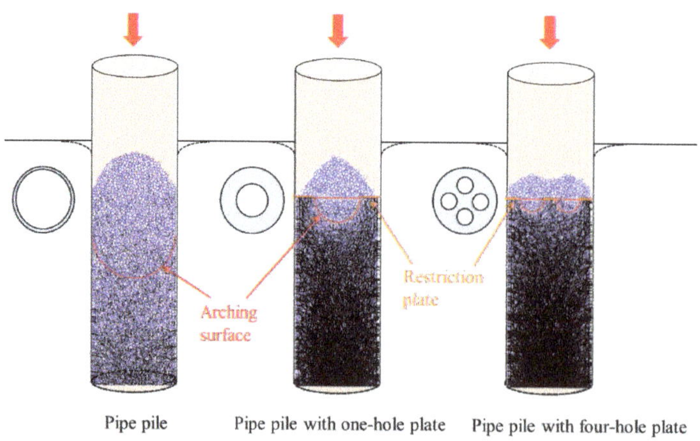

Figure 12. Force transmission and formation of soil arching inside soil plug.

6. Conclusions

In this paper, the current design practice for LDOEPPs was reviewed by comparing a few design methods used by different international associations. An experimental program, including laboratory-scale and geotechnical centrifuge pile load tests, was conducted to study the load-carrying behavior of pipe piles covering a wide range of pile diameters and embedment depths. The results were used to analyze the influence of soil plugging and the bearing behavior of LDOEPPs. Two innovatively designed restriction plates, i.e., a one-hole plate and a four-hole plate, were implemented inside the pipe piles and their effectiveness in facilitating the soil arching discussed. With certain simplifications and assumptions for the soil plug coefficient, a new method was proposed for a quick estimation of the bearing capacity of LDOEPPs with different restriction plates.

By comparing the normalized bearing capacity of pipe piles with different pile diameters, it was shown that small-diameter pipe piles tend to form a high degree of soil plugging, compared to large-diameter pipe piles. By using a restriction plate, the soil plug can be forced to form inside the pipe pile, which was proven by the laboratory-scale and centrifuge models. A four-hole restriction plate mobilizes more end bearing resistance than a one-hole restriction plate with the same plate opening area. While a higher end bearing capacity may cause drivability issues in the pile construction, the selection of one-hole and four-hole plates should be assessed on a case-by-case basis with considerations of mechanical performance and pile drivability. The experimental results were synthetized into a new method for bearing capacity evaluation by introducing a soil plug coefficient, which accounted for the effects of the soil plug as well as two different types of restriction plates. Overall, this study provides some insights into the behaviors of LDOEPPs. The integrated laboratory and geotechnical centrifuge experiments help to explain the formation of soil plugging. Due to schedule and resource constraints, the coverage of this study was not exhaustive. The design recommendations to account for the soil plug effect need to be further validated with the practice data, and other factors, e.g., pile corrosion, should be addressed in future study.

Author Contributions: Conceptualization, X.Y., Y.G. and J.L.; methodology and investigation, Y.G. and J.L.; writing—review and editing, Y.G. All authors have read and agreed to the published version of the manuscript.

Funding: The research project is supported by the Nature Science Foundation of Guangdong Province (Grant No. 2022A1515010693).

Institutional Review Board Statement: Not applicable.

Informed Consent Statement: Not applicable.

Data Availability Statement: Not applicable.

Conflicts of Interest: The authors declare no conflict of interest.

Appendix A

Table A1. Summary of laboratory and centrifuge experimental results.

	Pile No.	Prototype Diameter	Gravitational Acceleration	Prototype Embedment Depth	Ultimate Resistance, Pipe Pile	Ultimate Resistance, One-Hole Plate	Ultimate Resistance, Four-Hole Plate	Equiv. Tip Resistance, Pipe Pile	Equiv. Tip Resistance, One-Hole Plate	Equiv. Tip Resistance, Four-Hole Plate	Terzaghi Bearing Capacity	Tip Resistance Increment, One-Hole Plate	Tip Resistance Increment, Four-Hole Plate
		m		m	N	N	N	kPa	kPa	kPa	kPa		
Centrifuge Test	10, 11, 12	0.019	1-g	0.08	37.5	75	100	127.07	259.33	347.51	43.7	104.10%	173.50%
	10, 11, 12	0.19	10-g	0.8	7500	13,500	21,000	212.60	424.22	688.75	436.6	99.50%	224.00%
	10, 11, 12	0.57	30-g	2.4	140,000	215,000	285,000	392.88	686.80	961.12	1309.9	74.80%	144.60%
	10, 11, 12	0.95	50-g	4	670,000	890,000	1,080,000	685.64	996.01	1264.06	2183.1	45.30%	84.40%
	13, 14, 15	0.0254	1-g	0.08	125	190	270	242.81	371.09	528.97	44.4	52.80%	117.90%
	13, 14, 15	0.254	10-g	0.8	21,000	24,000	39,000	375.60	434.81	730.84	443.9	15.80%	94.60%
	13, 14, 15	0.762	30-g	2.4	400,000	500,000	610,000	760.61	979.89	1221.10	1331.7	28.80%	60.50%
	13, 14, 15	1.27	50-g	4	900,000	1,800,000	2,400,000	516.28	1226.75	1700.40	2219.5	137.60%	229.40%
	16, 17, 18	0.0384	1-g	0.08	350	520	650	299.65	446.44	558.69	45.9	49.00%	86.40%
	16, 17, 18	0.384	10-g	0.8	31,000	105,000	117,000	241.99	880.96	984.57	458.7	264.10%	306.90%
	16, 17, 18	1.152	30-g	2.4	390,000	1,375,000	1,700,000	297.10	1242.12	1553.93	1376.1	318.10%	423.00%
	16, 17, 18	1.92	50-g	4	2,740,000	4,300,000	5,400,000	817.92	1356.72	1736.65	2293.6	65.90%	112.30%
Lab Test	1, 2, 3	0.05078	1-g	0.45	1261	1421	1474	561.18	640.18	666.35	239.2	14.10%	18.70%
	4, 5, 6	0.09909	1-g	0.45	2929	4211	6542	348.31	514.56	816.82	244.7	47.70%	134.50%
	7, 8, 9	0.14876	1-g	0.45	4421	6254	8916	233.38	338.85	492.01	250.4	45.20%	110.80%

References

1. National Cooperative Highway Research Program (NCHRP). Design and Load Testing of Large-Diameter Open-Ended Driven Piles. In *NCHRP Synthesis 478*; Transportation Research Board: Washington, DC, USA, 2015.
2. Mohan, S.; Stevens, R.; Howard, R.; Kapuskar, M. Design of Large-Diameter Driven Pipe Pile Foundations: New East Span San Francisco-Oakland Bay Bridge. In Proceedings of the GeoTrans 2004, Los Angeles, CA, USA, 27–31 July 2004; pp. 278–285.
3. Doherty, P.; Gavin, K. Shaft Capacity of Open-Ended Piles in Clay. *J. Geotech. Geoenviron. Eng.* **2011**, *137*, 1090–1102. [CrossRef]
4. Stevens, R.F.; Wiltsie, E.A.; Turton, T.H. Evaluating Pile Drivability for Hard Clay, Very Dense Sand, and Rock. In Proceedings of the Fourteenth Annual Offshore Technology Conference, Houston, TX, USA, 3–6 May 1982; pp. 465–481.
5. Li, S.; Tian, Y.; Zhou, Y.; Jiang, B.; Wang, J. Premature Refusal of Large-Diameter, Deep-Penetration Piles on an Offshore Platform. *Appl. Ocean. Res.* **2013**, *42*, 55–59.
6. Liebich, B.A. High-Capacity Piles at the Stony Creek Bridge Project. *Transp. Res. Rec. J. Transp. Res. Board* **2009**, *2116*, 41–46. [CrossRef]
7. Kishida, H.; Isemoto, N. Behavior of Sand Plugs in Open-End Steel Pipe Piles. In Proceedings of the 9th International Society for Soil Mechanics and Foundation Engineering, Tokyo, Japan, 10–15 July 1977; pp. 601–604.
8. Randolph, M.F.; Dolwin, J.; Beck, R. Design of Driven Piles in Sands. *Geotechnique* **1994**, *44*, 427–448. [CrossRef]
9. Tomlinson, M.; Woodward, J. *Pile Design and Construction Practice*; Taylor & Francis: Oxfordshire, UK, 2008.
10. Paikowsky, S.; Whitman, R.V. The Effects of Plugging on Pile Performance and Design. *Can. Geotech. J.* **1990**, *27*, 429–440. [CrossRef]
11. Bradshaw, A.S.; Haffke, S.; Baxter, C. Load Transfer Curves from a Large-Diameter Pipe Pile in Silty Soil. In Proceedings of the GeoCongress 2012, Oakland, CA, USA, 25–29 March 2012; pp. 590–601.
12. Miller, G.A.; Lutenegger, A.J. Influence of Pile Plugging on Skin Friction in Overconsolidated Clay. *J. Geotech. Geoenviron. Eng.* **1997**, *123*, 525–533. [CrossRef]
13. Paik, K.; Salgado, R.; Lee, J.; Kim, B. Behavior of Open- and Closed-Ended Piles Driven into Sands. *J. Geotech. Geoenviron. Eng.* **2003**, *129*, 296–306. [CrossRef]
14. Gudavalli, S.R.; Safaqah, O.; Seo, H. Effect of Soil Plugging on Axial Capacity of Open-Ended Pipe Piles in Sands. In Proceedings of the 18th International Conference on Soil Mechanics and Geotechnical Engineering, Paris, France, 2–6 September 2013; pp. 1487–1490.
15. Dijkstra, J.; Broere, W. Experimental Investigation into Plugging of Open-Ended Piles. In Proceedings of the ASME 2009 28th International Conference on Ocean, Offshore and Arctic Engineering, Honolulu, HI, USA, 31 May–5 June 2009; pp. 1–8.
16. O'Neill, M.W.; Raines, R.D. Load Transfer for Pipe Piles in Highly Pressured Dense Sand. *J. Geotech. Eng.* **1991**, *117*, 1208–1226. [CrossRef]
17. Lehane, B.M.; Gavin, K.G. Base Resistance of Jacked Piles in Sand. *J. Geotech. Geoenvironmental Eng.* **2001**, *127*, 473–480. [CrossRef]
18. Randolph, M.F.; Leong, E.C.; Houlsby, G.T. One-Dimensional Analysis of Soil Plugs in Pipe Piles. *Geotechnique* **1991**, *41*, 587–598. [CrossRef]
19. Guo, Y.; Yu, X. Soil Plugging Mechanism on Large-Diameter Pipe Piles: Insight from Discrete Element Simulations. In Proceedings of the IFCEE 2015, San Antonio, TX, USA, 17–21 March 2015.
20. Zheng, C.; Liu, H.; Ding, X. Lateral Dynamic Response of a Pipe Pile in Saturated Soil Layer. *Int. J. Numer. Anal. Methods Geomech.* **2016**, *40*, 159–184. [CrossRef]
21. Peiffer, H. Predicting the Settlement Behaviour of Open-Ended Pipe Piles with a Large diameter. *Geotech. Lett.* **2017**, *7*, 225–229. [CrossRef]
22. Randolph, M.F.; Leong, E.C.; Hyden, A.M.; Murff, J.D. Soil Plug Response in Open-Ended Pipe Piles. *J. Geotech. Eng.* **1992**, *118*, 743–759. [CrossRef]
23. Liyanapathirana, D.S.; Deeks, A.J.; Randolph, M.F. Numerical Analysis of Soil Plug Behavior Inside Open-Ended Piles during Driving. *Int. J. Numer. Anal. Methods Geomech.* **1998**, *22*, 303–322. [CrossRef]
24. Gavin, K.; Lehane, B.M. The Shaft Capacity of Pipe Piles in Sand. *Can. Geotech. J.* **2003**, *40*, 36–45. [CrossRef]
25. McVay, M.C.; Badri, D.; Hu, Z.Y. *Determination of Axial Pile Capacity of Prestressed Concrete Cylinder Piles*; Report No. 4910-4504-877-12; Florida Department of Transportation: Tallahassee, FL, USA, 2004.
26. Paik, K.; Salgado, R. Determination of Bearing Capacity of Open-Ended Piles in Sand. *J. Geotech. Geoenviron. Eng.* **2003**, *129*, 46–57. [CrossRef]
27. Zvirko, O.I. In-Service Degradation of Structural Steels (A Survey). *Mater. Sci.* **2021**, *57*, 28–40. [CrossRef]
28. Kumara, J.J.; Kikuchi, Y.; Kurashina, T. Effects of the Lateral Stress on the Inner Frictional Resistance of Pipe Piles Driven into Sand. *Int. J. Geo-Eng.* **2016**, *7*, 1. [CrossRef]
29. Kien, H.P.; Ito, D. Bearing Capacity Characteristics of Steel Pipe Piles with Wings installed in Soil Cement Column. In Proceedings of the 6th International Conference of Euro Asia Civil Engineering Forum (EACEF 2017), Seoul, Korea, 22–25 August 2017.
30. Semple, R.S.; Gemeinhardt, J.P. Stress History Approach to Analysis of Soil Resistance to Pile Driving. In Proceedings of the Offshore Technology Conference, Houston, TX, USA, 4–7 May 1981; p. OTC-3869-MS.
31. American Petroleum Institute (API). ANSI/API Recommended Practice 2GEO. In *Geotechnical and Foundation Design Considerations*; American Petroleum Institute: Washington, DC, USA, 2011.
32. White, D.J.; Bolton, M.D. Comparing CPT and Pile Base Resistance in Sand. *Geotech. Eng.* **2005**, *158*, 3–14. [CrossRef]

33. Schneider, J.A.; Xu, X.; Lehane, B.M. End Bearing Formulation for CPT based Driven Pile. In Proceedings of the CPT'10 2nd International Symposium on Cone Penetration Testing, Huntington Beach, CA, USA, 9–11 May 2010; pp. 1–8.
34. Igoe, D.J.P.; Gavin, K.G.; O'Kelly, B.C. Shaft Capacity of Open-Ended Piles in Sand. *J. Geotech. Geoenviron. Eng.* **2011**, *137*, 903–913. [CrossRef]
35. Jeong, S.; Ko, J.; Won, J.; Lee, K. Bearing Capacity Analysis of Open-Ended Piles Considering the Degree of Soil Plugging. *Soils Found.* **2015**, *55*, 1001–1014. [CrossRef]
36. Yu, F.; Yang, J. Base Capacity of Open-Ended Steel Pipe Piles in Sand. *J. Geotech. Geoenviron. Eng.* **2012**, *138*, 1116–1128. [CrossRef]
37. Guo, Y.; Yu, X. Design and Analyses of Open-Ended Pipe Piles in Cohesionless Soils. *Front. Struct. Civ. Eng.* **2016**, *10*, 22–29. [CrossRef]
38. Hannigan, P.J.; Goble, G.G.; Likins, G.E.; Rausche, F. *Design and Construction of Driven Pile Foundations*; Federal Highway Administration: Washington, DC, USA, 2006.
39. Aas, P.; Karlsrud, K.; Clausen, C. Bearing Capacity of Driven Piles in Sand, the NGI Approach. In Proceedings of the Frontiers in Offshore Geotechnics: International Symposium on Frontiers in Offshore Geotechnics, Perth, WA, Australia, 19–21 September 2005; pp. 677–682.
40. *Steel Pipe Piles*; Finnish National Road Administration (FinnRA): Helsinki, Finland, 2000.
41. *JGJ94-2008*; China Ministry of Construction (CMC). Technical Code for Building Pile Foundations: Beijing, China, 2008.
42. Jardine, R.; Chow, F.; Overy, R.; Standing, J. *ICP Design Methods for Driven Piles in Sands and Clays*; Thomas Telford Publishing: London, UK, 2005.
43. Lehane, B.M.; Schneider, J.A.; Xu, X. *A Review of Design Methods for Offshore Driven Piles in Siliceous Sand*; The University of Western Australia: Perth, Australia, 2005.
44. Abu-Farsakh, M.Y.; Haque, M.N.; Tsai, C. A Full-Scale Field Study for Performance Evaluation of Axially Loaded Large-Diameter Cylinder Piles with Pipe Piles and PSC piles. *Acta Geotech.* **2017**, *12*, 753–772. [CrossRef]
45. Dong, J.; Chen, F.; Zhou, M.; Zhou, X. Numerical Analysis of the Boundary Effect in Model Tests for Single Pile under Lateral Load. *Bull. Eng. Geol. Environ.* **2018**, *77*, 1057–1068. [CrossRef]
46. *ASTM D1143/D1143M*; Standard Test Methods for Deep Foundations Under Static Axial Compressive Load, ASTM. ASTM International: West Conshohocken, PA, USA, 2013; pp. 1–15.
47. Wang, X.; Yang, X.; Zeng, X. Centrifuge Modeling of Lateral Bearing Behavior of Offshore Wind Turbine with Suction Bucket Foundation in Sand. *Ocean. Eng.* **2017**, *139*, 140–151. [CrossRef]
48. Iai, S.; Tobita, T.; Nakahara, T. Generalised Scaling Relations for Dynamic Centrifuge Tests. *Geotechnique* **2005**, *55*, 355–362. [CrossRef]
49. Truong, P.; Lehane, B.M.; Zania, V.; Klinkvort, R.T. Empirical Approach based on Centrifuge Testing for Cyclic Deformations of Laterally Loaded Piles in Sand. *Geotechnique* **2019**, *69*, 133–145. [CrossRef]
50. Li, Y.; Zhang, G.; Liu, C.Y. Effects of Pile Installation Simulation on Behavior of Pile Groups in Centrifuge Model Tests. *Geotech. Test. J.* **2018**, *41*, 815–820. [CrossRef]
51. *ASTM D2487*; ASTM International. Standard Practice for Classification of Soils for Engineering Purposes (Unified Soil 459 Classification System). ASTM International: West Conshohocken, PA, USA, 2011; pp. 1–10.
52. *NAVFAC DM 7.2*; Foundation and Earth Structures. U.S. Department of the Navy: Monterey, CA, USA, 1984.

sustainability

Article

Performance Investigation of Geopolymer Grouting Material with Varied Mix Proportions

Jianwei Liu [1,2], Han Feng [3], Yongxing Zhang [1,2,*] and Kaiqi Zheng [1,2]

1 School of Civil Engineering, Nanjing Forestry University, Nanjing 210037, China
2 Key Laboratory of Concrete and Pre-Stressed Concrete Structure of Ministry of Education, Southeast University, Nanjing 210096, China
3 School of Civil Engineering, Central South University, Changsha 410075, China
* Correspondence: zhanguongxing81@aliyun.com

Abstract: Grouting materials require not only high ultra-early-stage strength of the stone body, but also proper working performances, such as high fluidity and fast setting time, as well as good stability. Compared with the traditional pure cement grout, geopolymer grouting material has the advantages of fast setting time, high fluidity, good slurry stability, and high early strength of the slurry stone body, which is beneficial to reuse solid waste resources and can be applied to the conditions of rapid construction or repair work to a certain extent. This paper presents an experimental investigation into the performance variation of geopolymer grouting material with varied mass fractions of raw materials, and the grouting performance of geopolymer material with optimal mix proportion is also presented. The study is implemented by employing the designed experimental schemes, focused on fluidity and setting times, as well as ultra-early-stage (4 h, 8 h, 12 h, 16 h, 20 h, and 24 h) mechanical strength. The experimental result demonstrates that both ground granulated blast-furnace slag (GGBS) content and the mass ratio of activator solution to solid have influence on the working performance of geopolymers, and both GGBS content and activator concentration have influence on the mechanical strength of geopolymers. Furthermore, the variance analysis demonstrates that the fluidity of geopolymer material is dominantly affected by the mass ratio of activator solution to solid, the setting time of geopolymer material is mainly influenced by GGBS content, and the mechanical strength of geopolymer material is mainly affected by activator concentration. Moreover, the recommended mix proportion of geopolymer grouting material is proposed in this study, in which the replacement rate of GGBS is 45%; the modulus and concentration of modified sodium silicate activator are 1.5 and 75%, respectively; and the mass ratio of activator solution to solid is 1.5. In the recommended mix proportion, the geopolymer material has excellent comprehensive performance to implement grouting operation, in which the compressive and tensile strengths of the stone body reach 12.2 MPa and 0.8 MPa in 4 h, and reach 21.2 MPa and 2.1 MPa in 24 h. The fluidity is 223 mm, the initial setting and final setting times are 50 min and 57 min, the slurry stability of geopolymer material is good without liquid precipitation, in which the setting time is far less than 2 h. This work provides the experimental foundations for investigating the performance of geopolymer grouting material, which is also expected to provide reference for the further application and promotion of geopolymer materials used for grouting operations in rapid construction or repair work.

Keywords: geopolymer material; varied mix proportion; working performance; ultra-early-stage mechanical strength; experimental investigation

1. Introduction

Grouting materials require not only high ultra-early-stage strength of the stone body, but also proper working performances, such as high fluidity and fast setting time, as well as good stability [1]. In comparison to conventional cementitious grouting materials, geopolymer is an excellent grouting material with obvious advantages of fast hardening and high early strength, as well as corrosion resistance and high temperature resistance [2–4], which is beneficial to reuse solid waste resources and can be adopted in infrastructure construction, such as tunneling work [5–8]. Moreover, geopolymer is a kind of inorganic cementitious material, developed from raw materials with rich aluminosilicate, as well as synthesized by the action of an alkali activator, in which the curing reaction of geopolymer is different from the hydration of portland cement. As mentioned above, the application of geopolymer material as grouting material also requires high strength slurry stone body as well as proper working performance, and the factors that affect the performance of geopolymer material are complex and diverse. Among them, the type of raw materials adopted in mix proportions is one of the main influencing factors, in which the commonly used raw materials include metakaolin, fly ash, and blast furnace slag at present, and the potential pozzolanic activities of different kinds of raw materials are quite different. In view of previous investigations, the alkali activator and ground granulated blast-furnace slag (GGBS), as well as metakaolin (MK), are dominant compositions of geopolymer at present [9–12]. The commonly used metakaolin and blast furnace slag have higher potential pozzolanic activities than fly ash [13,14], in which the content of Si and Al in metakaolin is higher as well as conducive to the polymerization reaction [14], and blast furnace slag contains some calcium ions as well as being conducive to the early strength of stones at the same time. Moreover, another major factor affecting the performance of geopolymer is the composition of the alkali activator, in which many alkaline solutions can realize alkali excitation reaction, whereas NaOH solution and water glass composite solution are commonly used as the alkali activator in consideration of economic benefits, process conditions, and actual effects. The existing literature has studied the influence of the alkali activator on the performance of geopolymer material. The systematical investigation of the content of sodium hydroxide, sodium silicate, and water in the alkali activator, influencing the mechanical properties of geopolymer material, has been realized, in which the mechanical properties of geopolymer material is firstly increased and then decreased with the increase of alkali equivalent [15,16].

As described above, the mechanical strength and its growth rate as well as the working performance of setting time and fluidity are the main influencing factors of geopolymer material used for grouting application. In view of previous literatures, several researchers have carried out relevant studies from different perspectives on using the advantages of geopolymer materials for grouting operation in the rapid construction or repairing work [17–19], which is expected to obtain geopolymer grouting materials with suitable working performance and high physical strength. The existing research demonstrate that the factors affecting the mechanical properties of geopolymer material have been studied to a certain extent, and the application of geopolymer as grouting material has also been explored accordingly. Especially, the aforementioned mechanical strength and working performances of geopolymer material are varied with different types of raw materials in mix proportions [9,20–22], such as the very high activator concentration, which can easily induce the flash coagulation of geopolymer, and the material performance can show different trends with the increased amount of activator in the range of different alkali excitation dosage [23–25]. However, there are relatively few factors affecting the performance in the current research of geopolymer-based grouting material unfortunately, and there is no corresponding research on ultra-early-stage mechanical performance of geopolymer material used for the rapid grouting reinforcement of broken rock mass within a few hours, at present. Therefore, the performance of geopolymer grouting material with varied mix proportions has not yet been clearly understood until now, which may lead to an incomplete performance of geopolymer material in grouting operation, and there is therefore, an urgent requirement for the further study of the performance of geopolymer material used for grouting reinforcement.

In this paper, the experimental investigation into the ultra-early-stage mechanical strength and the working performances of fluidity and setting times in geopolymer material is implemented with varied mass fractions of raw materials in mix proportions. The study focuses on the performance variation of geopolymer materials influenced by varied mass fractions of raw materials with different mix proportions [26–29], considering the influence of four factors of ground granulated blast-furnace slag (GGBS). Sodium silicate modulus, activator concentration, as well as the mass ratio of activator solution to solid in the different mix proportions [30–32], in which the metakaolin mixed with slag is used as raw material, sodium hydroxide, and sodium silicate composite solution are used as activators. The optimum mix proportion of geopolymer with excellent material properties is recommended and used for grouting operation in the process of rapid construction or repair work, since high mechanical performance within a few hours and the excellent working performance of geopolymer grouting material are required in the rapid grouting and reinforcement of broken rock mass. The study promotes the comprehensiveness of geopolymer performance and application research, and further explores the performance potential of geopolymers in the early stage. Additionally, the possibility of using geopolymer as a rapid grouting material for strengthening broken rock mass with excellent working performance, in which the slurry fluidity, viscosity, setting time, and water separation rate and other working properties, as well as ultra-early-stage (4 h, 8 h, 12 h, 16 h, 20 h, and 24 h) mechanical properties, including compressive strength and tensile strength, are studied through experimental tests.

2. Experiment Design of Geopolymer Material with Varied Mass Fractions of Raw Materials in Mix Proportions

The experiment is designed to investigate the performance variation of geopolymer material influenced by varied mass fractions of raw materials in different mix proportions, in which the geopolymer is made of dry powders mixed with alkali-activated solution.

2.1. Selected Raw Materials in Geopolymer

Figure 1 demonstrates the typical raw materials selected in the mix proportions, in which the dry powders are composed of metakaolin (MK) and ground granulated blast-furnace slag (GGBS). Table 1 shows the ingredients and mass fractions of the used compositions, GGBS and MK, in which GGBS is S95 level and MK is produced by Hunan Chaopai Technology Co., Ltd. Figure 2 demonstrates the granularity distribution of cumulative volume content versus particle size in the compositions of GGBS and MK. Moreover, the alkali-activator is made from the mixed solutions of sodium hydroxide and sodium silicate, in which the sodium hydroxide solution is produced using water and tablets with analytical purity, and the sodium silicate solution has 34% concentration with 3.3 molar ratio of SiO_2 to Na_2O.

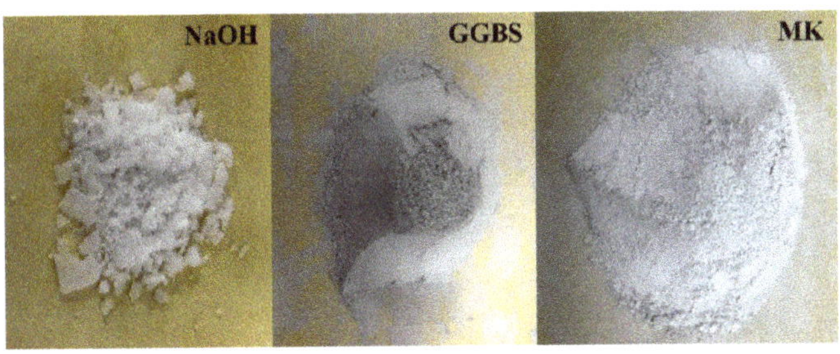

Figure 1. Typical raw materials selected in geopolymer material.

Table 1. Ingredients of metakaolin and ground granulated blast-furnace slag.

	Mass Fraction of Compositions (%)								
MK	SiO_2	Al_2O_3	Na_2O	TiO_2	K_2O	MgO	Fe_2O_3	CaO	LOI
	44.29	37.22	0.62	0.5	0.13	0.14	0.36	0.39	14.98
GGBS	CaO	SiO_2	Al_2O_3	MgO	SO_3	TiO_2	MnO	Fe_2O_3	K_2O
	36.91	33.98	15.22	9.27	1.81	0.80	0.59	0.62	0.41

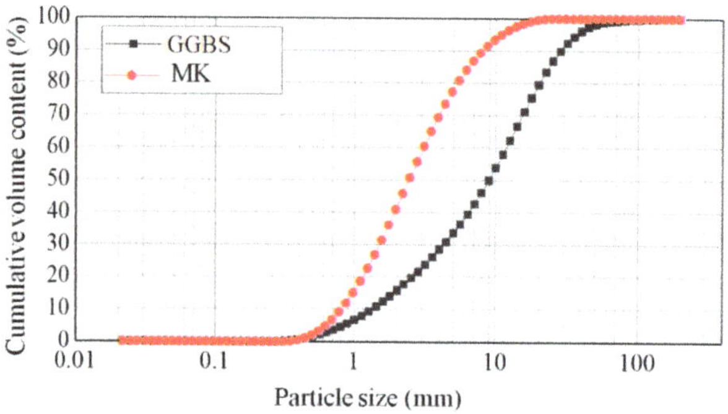

Figure 2. Granularity distribution in the compositions of MK and GGBS.

2.2. Design of Experimental Scheme

Table 2 demonstrates the designed schemes of experimental investigation into the performance of geopolymer material with varied mass fractions of raw materials in mix proportions, including GGBS content (A), sodium silicate modulus (B), activator concentration (C), as well as the mass ratio of activator solution to solid (D), in which B and C are molar ratio of SiO_2 to Na_2O in sodium silicate solution and the mass ratio of modified sodium silicate solution to activator solution, respectively. The sodium silicate modulus (B) can be changed by adding NaOH flakes with analytically pure levels in the solution, and activator concentration (C) is varied by adjusting water in the mix proportion and keeping for 12 h thereafter. Moreover, variance analysis is implemented to clarify the experimental result, in which the variance is the difference between maximum and minimum values of the statistical data [32]. The three levels of GGBS content (A) are 25%, 35%, and 45%, respectively; those of sodium silicate modulus (B) are 1.3, 1.4, and 1.5, respectively; those of activator concentration (C) are 55%, 65%, and 75%, respectively, and those of the mass ratio of activator solution to solid (D) are 1.4, 1.5, and 1.6, respectively, which are determined with referencing to previously implemented tests.

Table 2. Designed schemes of experimental investigation.

Scheme	Factor Compositions			
	A	B	C	D
1	25%	1.3	55%	1.4
2	25%	1.4	65%	1.5
3	25%	1.5	75%	1.6
4	35%	1.3	65%	1.6
5	35%	1.4	75%	1.4
6	35%	1.5	55%	1.5
7	45%	1.3	75%	1.5
8	45%	1.4	55%	1.6
9	45%	1.5	65%	1.4

2.3. Setup and Specimen Characteristics

Figure 3 shows the setup of a truncated circular die and a VICAT instrument, which were respectively adopted to test the fluidity and setting time of geopolymer material with varied mass fractions in mix proportion. The setting time of geopolymer material is recorded by VICAT instrument, and the fluidity of geopolymer material is determined as the maximum diameter of geopolymer material freely flowing on a glass plane over 30 s, which is tested by lifting the truncated circular die after fully filling with stirred grout.

Figure 3. Setup of truncated circular die and VICAT instrument.

Moreover, the compressive and flexural strengths of the stone body are also tested by a compression tester and a bending tester, in which all the specimens are cured in air condition with 20 ± 2 °C temperature. The bleeding rate of the geopolymer is determined using the measuring cylinder with 1000 mL capacity, in which the mixed batch of geopolymer is poured into the measuring cylinder with 850 mL volume. The initial height of geopolymer and height after keeping stationary for 2 h are recorded during the testing process, and the bleeding rate of geopolymer is accurately calculated by the following equation:

$$B = \frac{h_0 - h_1}{h_1} \tag{1}$$

where B is the bleeding rate, h_0 and h_1 are the initial height and height after keeping stationary for 2 h.

3. Performance Variation of Geopolymer Material with Varied Mix Proportions

3.1. Test Results

In the performance requirements of grouting material for rapid grouting reinforcement of broken rock mass, the early mechanical strength and strength growth rate of the stone body are the most important indicators. In view of previous studies, some scholars have conducted relevant research from different perspectives on employing the advantages of rapid setting and early strength in geopolymer materials used for grouting operations. These were conducted to obtain grouting materials with an appropriate working performance and a high physical strength of stones, in which the obtained 1d maximum compressive strength of alkali-activated geopolymer slurry stone body was 5.53 MPa [33]. Table 3 shows the statistical analysis of compression and tensile strengths of slurry stone body from test results, in which 4 h, 8 h, 12 h, 16 h, 20 h, and 24 h represent 4 h, 8 h, 12 h, 16 h, 20 h, and 24 h; f_c and f_t are the compression and tensile strengths of geopolymer material, respectively. It can be seen that the maximum compressive and tensile strengths of the stone body after the solidification of geopolymer grout are 12.2 MPa and 0.8 MPa after 4 h, and the maximum compressive and tensile strengths are 21.2 MPa and 2.1 MPa after 24 h. As discussed above, the compressive strength of 4h slurry stone body obtained from this study can reach 12 MPa, which is obviously larger than the 1d maximum compressive strength of geopolymer slurry stone body with 5.53 MPa in the existing literature. Therefore, the mass fractions of raw materials in mix proportions proposed in this study have certain advantages from the perspective of early mechanical strength of slurry stone body.

Table 3. Statistical analysis of compression and flexure strengths of stone body from test results.

Scheme	4 h (MPa)		8 h (MPa)		12 h (MPa)		16 h (MPa)		20 h (MPa)		24 h (MPa)	
	f_c	f_t	f_c	f_t	f_c	f_t	f_c	f_t	f_c	f_t	f_c	f_t
1	1.8	0.5	2.4	0.5	1.7	0.6	2.5	0.6	2.3	0.7	2.2	1.2
2	4.4	0.2	6.1	0.6	6.1	0.9	7.4	1.3	7.6	1.3	9.6	1.6
3	5.1	0.8	7.2	0.9	8.5	1.2	10.8	1.6	12.2	1.8	13.4	2.0
4	4.2	0.7	5.3	1.1	6.0	1.2	7.1	1.4	9.3	1.6	10.8	1.7
5	7.6	0.7	9.5	0.9	10.2	1.2	14.2	1.5	16.8	2.0	18.6	2.3
6	1.4	0.3	1.8	0.5	2.1	0.6	2.2	0.8	3.5	0.9	3.8	1.0
7	12.2	0.8	14.1	1.2	18.0	1.7	19.5	1.9	20.8	2.0	21.2	2.1
8	0.5	0.5	0.9	0.6	1.4	0.7	2.0	0.8	3.4	1.2	4.1	1.4
9	6.2	1.1	8.9	1.2	10.1	1.6	11.8	1.6	12.7	1.7	14.1	1.8

Table 4 demonstrates the statistical analysis of slurry performance from test results. The good fluidity of the fresh grout is a necessary condition for grouting operation, and the maximum fluidity of the fresh grout is 231 mm. Moreover, the proper setting time represents the pumpability and timeliness of the slurry, and the proper time interval between the initial setting time and the final setting time is conducive to the subsequent construction process at the same time. It can be seen that the initial setting time of the slurry described in this paper ranges from 35 min to 62 min, the final setting time ranges from 46 min to 72 min, and the interval between the initial setting time and the final setting time ranges from 6 min to 15 min.

Table 4. Statistical analysis of slurry performance from test results.

Scheme	Factor Compositions				Fluidity (mm)	Setting Time (min)	
	A	B	C	D		Initial Setting	Final Setting
1	0.25	1.3	55%	1.4	201	54	63
2	0.25	1.4	65%	1.5	208	56	65
3	0.25	1.5	75%	1.6	215	59	66
4	0.35	1.3	65%	1.6	227	52	60
5	0.35	1.4	75%	1.4	212	48	54
6	0.35	1.5	55%	1.5	221	58	63
7	0.45	1.3	75%	1.5	223	50	57
8	0.45	1.4	55%	1.6	229	54	61
9	0.45	1.5	65%	1.4	213	46	54

Based on the statistical analysis of the aforementioned mechanical strength and working performance, the optimal mix proportion of the geopolymer material is determined by comprehensively considering the significance of each index, in which the replacement rate of mineral powder is 45%, the modulus and concentration of the modified sodium silicate activator are 1.5 and 75%, respectively, and the mass ratio of activator solution to solid is 1.5.

3.2. Working Performance Variation

As one of the main performance indexes of grouting operation during the process of rapid construction or repairing work, fluidity affects the diffusion ability of geopolymer grouting material and also relates to the difficulty of grouting operation. Figure 4 demonstrates the fluidity variation of geopolymer influenced by varied mass fractions of raw materials in mix proportions, in which the influence from varied mass fractions of GGBS content (A), sodium silicate modulus (B), activator concentration (C), and the mass ratio of activator solution to solid (D) are studied. It can be clearly seen that all the fluidity results of the tested geopolymer materials are larger than 200 mm, which are little changed with the increasing content of sodium silicate modulus (B) and activator concentration (C); whereas those tend to gradually increase with the increasing mass ratio of activator solution to solid (D). Moreover, the geopolymer fluidities are increased with the increasing GGBS content (A), whereas the fluidity increment is reduced thereafter, in which the adsorbability of composite powder is relatively decreased with the increase of GGBS content (A), the metakaolin (MK) itself has strong adsorbability, and the free water in the slurry increases with a certain mass ratio of activator solution to solid (D), and then the fluidity of the slurry is thus increased.

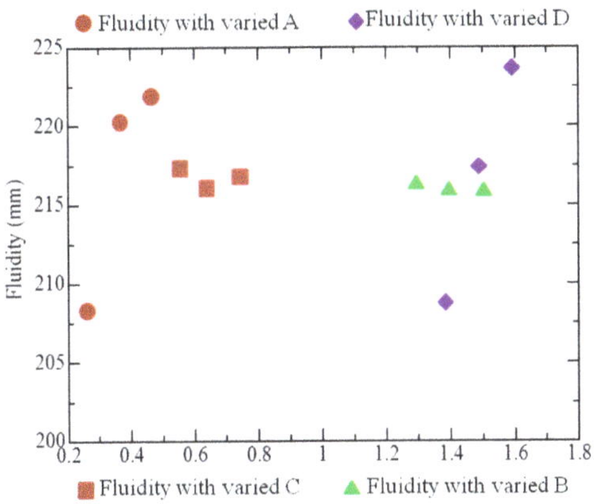

Figure 4. Geopolymer fluidity with varied mass fractions of raw materials in mix proportion.

Grout setting time is a key indicator of grouting operation in the process of rapid construction or repair work. Setting time that is too short is not conducive to field operation, which cannot guarantee the slurry diffusion range, since it is easy to cause blockage of grouting machines and tools. Through analysis of the initial setting time and final setting time of the slurry, the influence trend of each influencing factor on the setting time of the slurry is obtained. Figure 5 demonstrates the setting times variation in geopolymer with varied mass fractions of raw materials in mix proportions, in which the influence from varied mass fractions of GGBS content (A), sodium silicate modulus (B), activator concentration (C), and the mass ratio of activator solution to solid (D) are studied. It can be clearly seen geopolymer has the advantage of a short setting time far less than 2 h, in which the initial setting time is varied from 35 min to 62 min, and the final setting time is varied from 46 min to 72 min; the interval of both is ranged from 6 min to 15 min. There are few changes in both initial and final setting times with the increasing content of sodium silicate modulus (B) and activator concentration (C), whereas those are decreased with the increasing GGBS content (A). Especially, both initial and final setting times are first increased with the increasing mass ratio of activator solution to solid (D), whereas the time increment is reduced thereafter. This behavior is quite different from ordinary cement slurry. Moreover, the geopolymer has good stability without liquid precipitation in the experiment, which is probably due to the short setting time being far less than 2 h, as well as the water retaining capacity of metakaolin (MK) in geopolymer materials.

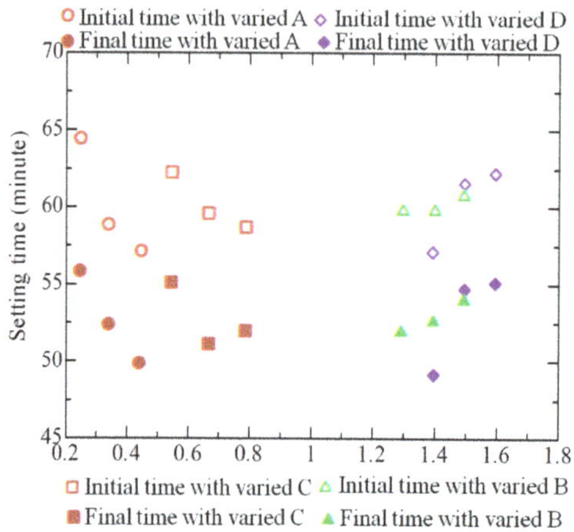

Figure 5. Setting times with varied mass fractions of raw materials in mix proportion.

The mineral powder contains calcium ions, which have a stronger charge attraction than sodium ions. Therefore, the initial setting time and final setting time of the slurry are reduced with increasing GGBS content (A). In view of previous literature, the theoretical reaction model of the geopolymer includes three processes: dissolution, recombination, and polycondensation [34]. The coagulation time is the performance of the reaction rate to a certain extent, in which the dissolution, recombination, and polycondensation processes overlap each other when the activator concentration (C) is low, and the gel is wrapped around the undissolved aluminosilicate, inhibiting the reaction and resulting in an increase in the coagulation time. On the contrary, the reaction has an obvious sequence and the dissolution and polycondensation processes are clearly different when the activator concentration (C) is high, in which the reaction rate is also faster and the coagulation time is reduced [34]. The increase in the mass ratio of activator solution to solid (D) separates the distance between solid particles, and the spatial flocculation structure can't be formed in time, which increases the coagulation time of geopolymer material.

This implies that both sodium silicate modulus (B) and activator concentration (C) have little effect on the working performance of geopolymer material. However, both GGBS content (A) and the mass ratio of activator solution to solid (D) have influence on the working performance of geopolymer material, in which the implemented variance analysis demonstrates that the fluidity is dominantly affected by the mass ratio of activator solution to solid (D), and the setting time is mainly influenced by GGBS content (A).

3.3. Mechanical Strength Variation

In the special requirements for rapid grouting reinforcement, the early mechanical strength of the slurry stone body is one of the most important indexes among all the performance indexes. Figure 6 demonstrates the compressive and tensile strengths in geopolymer material with varied mass fractions of raw materials in mix proportions, in which the influence from varied mass fractions of GGBS content (A), sodium silicate modulus (B), activator concentration (C), and the mass ratio of activator solution to solid (D) are studied. It can be clearly seen that there are few changes in the compressive and tensile strengths with the increasing content of sodium silicate modulus (B) and the mass ratio of activator solution to solid (D), whereas those are linearly increased with the increasing GGBS content (A) and activator concentration (C). This implies that both sodium silicate modulus (B) and the mass ratio of activator solution to solid (D) have little effect on the mechanical strength of geopolymer. However, both GGBS content (A) and activator concentration (C) have influence on the mechanical strength of geopolymer, in which the implemented variance analysis demonstrates that the mechanical strength is mainly affected by activator concentration (C) in the mix proportions.

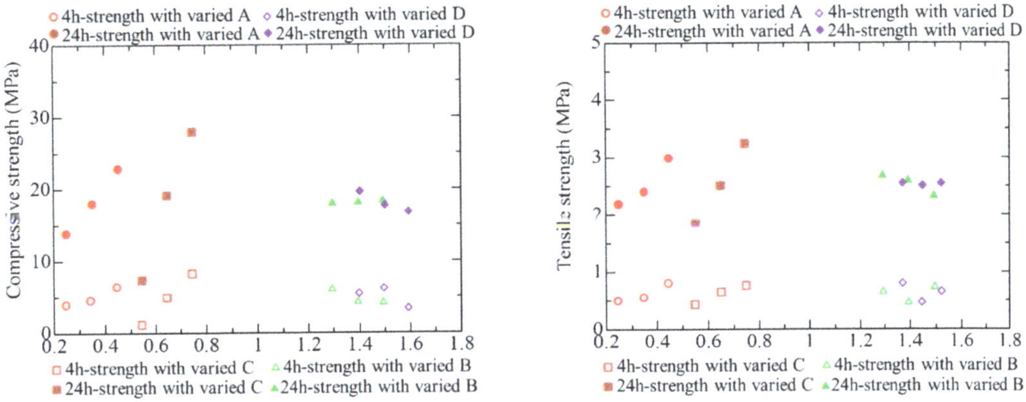

(**a**) (**b**)

Figure 6. Mechanical strength with varied mass fractions of raw materials in mix proportion. (**a**) Compressive strength variation. (**b**) Tensile strength variation.

4. Performance of Geopolymer Material with Recommended Mix Proportion

In view of the above experimental investigation, the aforementioned Scheme 7 in Table 2 is recommended as an optimum mix proportion of geopolymer material used for grouting operation in the process of rapid construction or repair work, in which GGBS content (A) and activator concentration (C) are 45% and 75%, respectively, sodium silicate modulus (B) is 1.3, and the mass ratio of activator solution to solid (D) is 1.5.

Figure 7 demonstrates the geopolymer material with recommended mix proportion, as well as the scanning electron microscope (SEM) image with 10,000 times, in which dense internal structure can be clearly observed in the geopolymer material with recommended mix proportion. Table 5 shows the working performance and mechanical strength of geopolymer with recommended mix proportion, in which the initial and final setting times are 49 min and 57 min, respectively, and the fluidity is 223 mm. This means the recommended geopolymer has good fluidity and an appropriate setting time for slurry diffusion in grouting operations. Moreover, the ultra-early-stage mechanical strength of the recommended geopolymer adequately satisfies the requirements of grouting reinforcement.

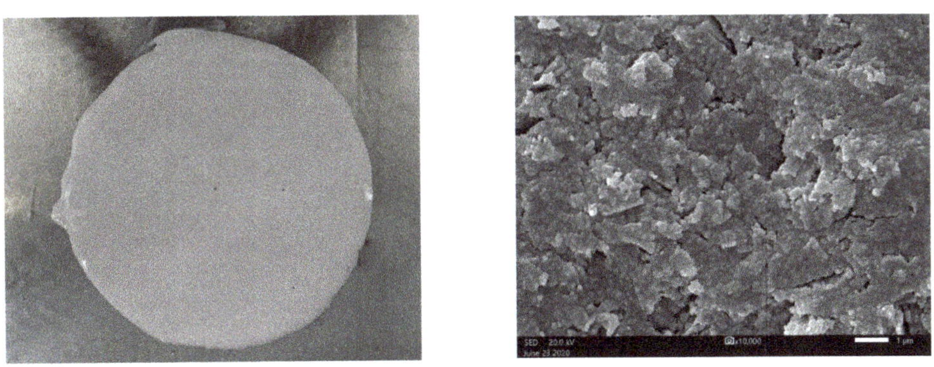

(**a**) (**b**)

Figure 7. Recommended geopolymer material and SEM image. (**a**) Geopolymer material with recommended mix proportion. (**b**) SEM image at 10,000 times.

Table 5. Working performance and mechanical strength of recommended geopolymer material.

Composition				Fluidity (mm)	Setting Time (min)		Strength (MPa)			
							Compressive		Tensile	
A	B	C	D		Initial	Final	4 h	24 h	4 h	24 h
45%	1.3	75%	1.5	223	49	57	12.2	21.2	0.8	2.1

The performance of the slurry stone body left by the grout with good stability can ensure good homogeneity, and the slurry stability is also one of the key factors to ensure the effect of grouting operation in the process of rapid construction or repair work, in which the water separation rate is generally taken account by an index for calibrating the stability of slurry, and the slurry is considered as stable if the water separation rate is less than 5% within 2 h. Different from the traditional cement slurry, the setting time of the geopolymer grouting material in this study is far less than 2 h, in which the metakaolin in the composition system of the geopolymer grouting material has the characteristics of strong water retention, as well as being similar to that of kaolin at the same time, and the recommended geopolymer grouting material thus has good stability in this study.

5. Conclusions

In this paper, the experimental investigation into the performance variation of geopolymer grouting material with varied raw materials is implemented by employing the designed experimental schemes, in which the applicability, ultra-early-stage working performance, and mechanical performance of geopolymer material used for grouting operation in the process of rapid construction or repairing work are studied through tests with varied mix proportions. Range analysis is adopted to investigate the specific working performance of the fresh grout and the physical strength of the slurry stone body, as well as the performance impact of geopolymer grouting material induced by the substitution rate of ground granulated blast-furnace slag (GGBS), sodium silicate modulus, activator concentration, and the mass ratio of activator solution to solid. Moreover, the recommended mix proportion of geopolymer grouting material is proposed, the grouting performance of geopolymer material with optimal mix proportion is also presented, and the following conclusions can be drawn.

(1) Both sodium silicate modulus and activator concentration have little effect on the working performance of geopolymer, whereas both GGBS content and the mass ratio of activator solution to solid have influence on the working performance of geopolymer, in which the variance analysis demonstrates that the fluidity is dominantly affected by the mass ratio of activator solution to solid, and the setting time is mainly influenced by GGBS content.

(2) Both sodium silicate modulus and the mass ratio of activator solution to solid have little effect on the mechanical strength of geopolymer, whereas both GGBS content and activator concentration have influence on the mechanical strength of geopolymer, in which the variance analysis demonstrates that the mechanical strength is mainly affected by activator concentration.

(3) The geopolymer material with the recommended mix proportion has good fluidity and appropriate setting time for slurry diffusion in grouting operations and the ultra-early-stage mechanical strength of the recommended geopolymer material adequately satisfies the requirement of grouting application, in which a dense internal structure can be observed in the image obtained by scanning electron microscope (SEM).

(4) Considering the working performance of the fresh grout and the physical strength of the slurry stone body, the optimal mass fractions of raw materials in mix proportion of geopolymer material used for grouting operation is suggested, in which the replacement rate of mineral powder is 45%, sodium silicate modulus and activator

concentration are 1.5 and 75%, respectively, and the mass ratio of activator solution to solid is 1.5.

(5) Compared with the traditional pure cement grout, the geopolymer grouting material has obvious advantages in rapid construction or repairing work, as well as green and environmental protection, mainly manifested in the strong early mechanical strength of the slurry stone body and excellent working performances of fluidity and setting time, as well as good stability without liquid precipitation, in which the setting time is far less than 2 h.

(6) In the recommended mix proportion, the geopolymer material has excellent comprehensive performance to implement grouting operation, in which the compressive and tensile strengths of the stone body reach 12.2 MPa and 0.8 MPa in 4 h, and those reach 21.2 MPa and 2.1 MPa in 24 h. The fluidity is 223 mm, the initial setting and final setting times are 50 min and 57 min, respectively, and the slurry stability of geopolymer material is good. Specifically, the aforementioned 4 h compressive strength of slurry stone body in the geopolymer grouting material with recommended mix proportion can reach 12.2 MPa, which has an obvious advantage over existing geopolymer material.

Author Contributions: Conceptualization, J.L., H.F. and Y.Z.; validation, J.L., H.F. and K.Z.; investigation, J.L. and H.F.; data curation, J.L. and H.F.; writing—original draft preparation, J.L.; writing—review and editing, H.F. and Y.Z.; supervision, Y.Z.; funding acquisition, J.L. and K.Z. All authors have read and agreed to the published version of the manuscript.

Funding: This research was funded by Natural Science Foundation of Jiangsu Province (No. BK20211281), Postgraduate Research and Practice Innovation Program of Jiangsu Province (KYCX22_1060), Six Talent Peak Projects and Open Fund Project from Key Laboratory of Concrete and Pre-stressed Concrete Structure of Ministry of Education (Southeast University) (CPCSME2018-10).

Institutional Review Board Statement: Not applicable.

Informed Consent Statement: Not applicable.

Data Availability Statement: The data used to support the findings of this study are available from the corresponding author upon request.

Conflicts of Interest: The authors declare that there are no conflicts of interest regarding the publication of this paper.

References

1. Zhang, C.; Yang, J.S.; Fu, J.Y.; Wang, S.Y.; Yin, J.; Xie, Y.P.; Li, L.Y. Cement based eco-grouting composite for pre-reinforcement of shallow underground excavation in vegetation protection area. *Tunn. Undergr. Space Technol.* **2021**, *118*, 104188. [CrossRef]
2. Atis, C.D.; Bilim, C.; ÇElik, Ö.; Karahan, O. Influence of activator on the strength and drying shrinkage of alkali-activated slag mortar. *Constr. Build. Mater.* **2009**, *23*, 548–555. [CrossRef]
3. Brough, A.R.; Atkinson, A. Sodium silicate-based, alkali-activated slag mortars: Part I. strength, hydration and microstructure. *Cem. Concr. Res.* **2002**, *32*, 865–879. [CrossRef]
4. Burciaga-DíAz, O.; Magallanes-Rivera, R.X.; Escalante-GarcíA, J.I. Alkali-activated slag-metakaolin pastes: STRENGTH, structural, and microstructural characterization. *J. Sustain. Cem. Based Mater.* **2013**, *2*, 111–127. [CrossRef]
5. Yao, Z.G.; Fang, Y.; Yu, T.; Pu, S.; Luo, H.; Cui, J.; Wang, J. Dynamic failure mechanism of tunnel under rapid unloading in jointed rockmass: A case study. *Eng. Fail. Anal.* **2022**, *141*, 106634. [CrossRef]
6. Yao, C.F.; He, C.; Huang, X.; Takemura, J.; Yang, W.B.; An, Z.L. Response of a continuous pipeline in sand subjected to normal faulting. *Transp. Geotech.* **2022**, *36*, 100824. [CrossRef]
7. He, C.; Feng, K.; Fang, Y.; Jiang, Y.C. Surface settlement caused by twin-parallel shield tunnelling in sandy cobble strata. *J. Zhejiang Univ. Sci. A* **2012**, *13*, 858–869. [CrossRef]
8. Wang, S.M.; Lin, Z.Y.; Peng, X.Y.; Wang, X.M.; Tu, G.; Song, Z.H. Research and evaluation on Water-dispersion resistance of synchronous grouting slurry in shield tunnel. *Tunn. Undergr. Space Technol.* **2022**, *129*, 104679. [CrossRef]
9. Kumar, S.; Kumar, R.; Mehrotra, S.P. Influence of granulated blast furnace slag on the reaction, structure and properties of fly ash based geopolymer. *J. Mater. Sci.* **2009**, *45*, 607–615. [CrossRef]
10. Provis, J.L.; Bernal, S.A. Geopolymers and related alkali-activated materials. *Annu. Rev. Mater. Res.* **2014**, *44*, 299–327. [CrossRef]

11. Vavričuk, A.; Bokan-Bosiljkov, V.; Kramar, S. The influence of metakaolin on the properties of natural hydraulic lime-based grouts for historic masonry repair. *Constr. Build. Mater.* **2018**, *172*, 706–716. [CrossRef]
12. Emdadi, Z.; Asim, N.; Amin, M.H.; Yarmo, M.A.; Maleki, A.; Azizi, M.; Sopian, K. Development of green geopolymer using agricultural and industrial waste materials with high water absorbency. *Appl. Sci.* **2017**, *7*, 514. [CrossRef]
13. Wang, M.; Jia, D.; He, P. Influence of calcination temperature of kaolin on the structure and properties of final geopolymer. *Mater. Lett.* **2010**, *64*, 2551–2554. [CrossRef]
14. Nadoushan, M.J.; Ramezanianpour, A.A. The effect of type and concentration of activator on flowability and compressive strength of natural pozzolan and slag-based geopolymers. *Constr. Build. Mater.* **2016**, *111*, 337–347. [CrossRef]
15. Peng, H.; Li, S.L.; Cai, C.S.; Zhang, X.F.; Cui, C. Study on Effect of Mix and Curing Conditions on the Mechanical Properties and Setting Time of Metakaolin-Based Geopolymer. *Bull. Chin. Ceram. Soc.* **2014**, *33*, 2809–2817.
16. Ghafoori, N.; Najimi, M.; Radke, B. Natural pozzolan-based geopolymers for sustainable construction. *Environ. Earth Sci.* **2016**, *75*, 1110. [CrossRef]
17. Guo, X.; Shi, H.; Dick, W.A. Compressive strength and microstructural characteristics of class C fly ash geopolymer. *Cem. Concr. Compos.* **2010**, *32*, 142–147. [CrossRef]
18. Soutsos, M.; Boyle, A.P.; Vinai, R.; Hadjierakleous, A.; Barnett, S.J. Factors influencing the compressive strength of fly ash based geopolymers. *Constr. Build. Mater.* **2016**, *110*, 355–368. [CrossRef]
19. Temuujin, J.; Riessen, A.V.; Mackenzie, K.J.D. Preparation and characterization of fly ash based geopolymer mortars. *Constr. Build. Mater.* **2010**, *24*, 1906–1910. [CrossRef]
20. Puligilla, S.; Mondal, P. Role of slag in microstructural development and hardening of fly ash-slag geopolymer. *Cem. Concr. Res.* **2013**, *43*, 70–80. [CrossRef]
21. Liu, Y.; Zhu, W.; Yang, E.H. Alkali-activated ground granulated blast-furnace slag incorporating incinerator fly ash as a potential binder. *Constr. Build. Mater.* **2016**, *112*, 1005–1012. [CrossRef]
22. Buchwald, A.; Hilbig, H.; Kaps, C. Alkali-activated metakaolin-slag blends-performance and structure in dependence of their composition. *J. Mater. Sci.* **2007**, *42*, 3024–3032. [CrossRef]
23. Heah, C.; Kamarudin, H.; Bakri, A.M.; Bnhussain, M.; Luqman, M.; Nizar, I.K.; Ruzaidi, C.M.; Liew, Y.M. Study on solids-to-liquid and alkaline activator ratios on kaolin-based geopolymers. *Constr. Build. Mater.* **2012**, *35*, 912–922. [CrossRef]
24. Alonso, S.; Palomo, A. Alkaline activation of metakaolin and calcium hydroxide mixtures: Influence of temperature, activator concentration and solids ratio. *Mater. Lett.* **2011**, *47*, 55–62. [CrossRef]
25. Alonso, S.; Palomo, A. Calorimetric study of alkaline activation of calcium hydroxide-metakaolin solid mixtures. *Cem. Concr. Res.* **2001**, *31*, 25–30. [CrossRef]
26. Kazea, C.R.; Djobob, J.N.Y.; Nanae, A. Effect of silicate modulus on the setting, mechanical strength and microstructure of iron-rich aluminosilicate (laterite) based-geopolymer cured at room temperature. *Ceram. Int.* **2018**, *44*, 21442–21450. [CrossRef]
27. Hadi, M.N.S.; Al-Azzawi, M.; Yu, T. Effects of fly ash characteristics and alkaline activator components on compressive strength of fly ash-based geopolymer mortar. *Constr. Build. Mater.* **2018**, *175*, 41–54. [CrossRef]
28. Robayo-Salazar, R.A.; GutiéRrez, R.M.D.; Puertas, F. Effect of metakaolin on natural volcanic pozzolan-based geopolymer cement. *Appl. Clay Sci.* **2016**, *132*, 491–497. [CrossRef]
29. Peng, H.; Ge, Y.P.; Cai, C.S.; Zhang, Y.X.; Liu, Z. Mechanical properties and microstructure of graphene oxide cement-based composites. *Constr. Build. Mater.* **2019**, *194*, 102–109. [CrossRef]
30. Morsy, M.; Alsayed, S.; Al-Salloum, Y.; Almusallam, T. Effect of sodium silicate to sodium hydroxide ratios on strength and microstructure of fly ash geopolymer binder. *Arab. J. Sci. Eng.* **2014**, *39*, 4333–4339. [CrossRef]
31. De Silva, P.; Sagoe-Crenstil, K.; Sirivivatnanon, V. Kinetics of geopolymerization: Role of Al_2O_3 and SiO_2. *Cem. Concr. Res.* **2007**, *37*, 512–518. [CrossRef]
32. Zou, G.; Xu, J.; Wu, C. Evaluation of factors that affect rutting resistance of asphalt mixes by orthogonal experiment design. *Int. J. Pavement Eng. Technol.* **2017**, *10*, 282–288. [CrossRef]
33. Wang, J.; Zhang, L.W.; Feng, X.; Zhao, S.L.; Wang, H.B. Me Experiment and application research on alkali-activated geopolymer two-component grouting material. *Chin. J. Rock Mech. Eng.* **2015**, *34*, 4418–4425. (In Chinese)
34. Peng, H.; Cui, C.; Cai, C.S.; Li, S.L.; Zhao, J.W. Mechanism of activator concentration influencing properties of metakaolin-based geopolymer. *Acta Mater. Compos. Sin.* **2016**, *33*, 2952–2960. (In Chinese)

sustainability MDPI

Article

A New p–y Curve for Laterally Loaded Large-Diameter Monopiles in Soft Clays

Mingyuan Wang [1], Miao Wang [2], Xinglei Cheng [2,*], Qun Lu [2] and Jiaqing Lu [2]

[1] Powerchina Huadong Engineering Corporation Limited, Hangzhou 310014, China
[2] Key Laboratory of Soft Soil Engineering Character and Engineering Environment of Tianjin, Tianjin Chengjian University, Tianjin 300384, China
* Correspondence: chengxinglei110@163.com

Abstract: In harsh offshore environmental conditions, the monopile foundations supporting offshore wind turbines must be designed for lateral loads such as winds, waves, and currents. The Beam on Nonlinear Winkler Foundation (BNWF) method has been widely used because of its clear concept and lower calculation cost. The selection of a reasonable p–y curve is critical to the calculation accuracy of this method. This paper clarified the defects of widely used API p–y curves for soft clays and then proposed a new p–y curve with better versatility and applicability. The suitability of the proposed p–y curve was validated by comparing it with the calculation results from the three-dimensional finite element method (3D FEM). Compared with the API p–y curve, the proposed p–y curve can better predict the lateral behavior of large-diameter piles in soft clays, such as the load–displacement curve of the pile head, lateral deflection profile, and bending moment profile. The research findings can provide guidance for the design of monopile foundations supporting offshore wind turbines in soft clays.

Keywords: p–y curve; lateral loading; large-diameter monopile; BNWF method; soft clays

1. Introduction

Natural resources such as coal, petroleum, and natural gas are being consumed at an alarming rate, hence many countries are actively promoting green and sustainable new energy technologies. Wind energy is one of the most promising new energies due to its advantages of cleanness and renewability. In addition, compared with land-based wind energy, offshore wind energy is attractive for its various advantages including high speed, low wind shear, low turbulence, and high output. The urgent demand for wind power is promoting the development of construction technology for offshore wind turbines (OWTs) [1]. The design of the foundations supporting OWTs is extremely important because it may account for as much as 30% of the total cost of a typical offshore wind project [2,3]. Nonetheless, the large diameter monopile foundation remains by far the most popular foundation type being deployed in more than 75% of the currently installed OWTs [4,5]. In harsh offshore environmental conditions, pile foundations must be designed for lateral loads because OWTs are always subjected to such loads caused by winds, waves, and currents.

The Beam on Nonlinear Winkler Foundation (BNWF) method based on the p–y curve is widely used to analyze the behavior of laterally loaded piles. McClelland and Focht [6] first proposed the concept of a p–y curve by comparing the measured curve of lateral soil resistance versus the pile lateral deflection with the soil stress–strain curve in the consolidated undrained tests. Matlock [7] proposed a p–y curve model for piles installed in clay based on limited pile load tests at the Sabine River site, which was adopted by the API RP 2GEO (American Petroleum Institute (API)) [8] and is widely used in industry. In the BNWF approach, the pile is simulated as an elastic beam while the soil is idealized as a series of discrete nonlinear springs along the depth of the pile. The nonlinear p–y springs describe the relationship between the local lateral soil resistance p and the pile lateral relative displacement y. The BNWF method based on the p–y curve is actually a simplified method ignoring the soil continuity.

In recent decades, some new p–y curves for soft clay have been proposed to analyze the lateral behavior of piles. The hyperbolic equation, initially proposed by Kondner [9] and later identified by Georgiadis et al. [10] as a suitable formulation, is used to define the shape of the p–y curves. Dewaikar and Patil et al. [11] developed an improved method for the construction of hyperbolic p–y curves, in which the initial stiffness of the p–y curve is taken to be increasing with depth. Jeanjean [12] proposed a hyperbolic tangent p–y curve and verified its rationality using the centrifugal test and finite element simulation results of pile in clays. Klinkvort and Hededal [13] performed a series of centrifuge tests to investigate the behavior of a rigid pile loaded with a high eccentricity and presented a reformulation of the p–y curve. Rathod et al. [14] developed new p–y curves for piles located on crests of soft clay with different sloping ground surfaces under static lateral loading by carrying out a series of laboratory model tests. Zhu et al. [15] modified the hyperbolic tangent p–y curve by introducing an additional constant and verified its rationality using the field tests of offshore driven piles. Xu et al. [16] found that the hyperbolic method yields larger prediction errors than the API method and proposed a modified p–y approach based on field measurements. The p–y curve method has been developed to better simulate lateral responses based on model tests and numerical simulation results [17–24]. The above research mainly focus on the p–y curve under lateral monotonic loads.

The pile cyclic lateral response was also analyzed by developing cyclic p–y hysteresis curves based on the Beam on Nonlinear Winkler Foundation (BNWF) method. In these models, some pile–soil interaction characteristics are simulated using specific components such as spring, dashpot, drag cell, and gap cell to capture the dynamic hysteretic behavior of the soil–pile interface [25–29]. However, the introduction of too many components usually makes the implementation of the model more complex. In addition to the p–y method, the three-dimensional finite element method (3D FEM) was also used to analyze the lateral response of piles even though it requires larger costs of calculation [30–38]. The 3D FEM can better simulate the pile–soil interaction because both the pile and soil are simulated by continuum elements. Compared with the method, although discrete the p–y springs do not rigorously capture the soil continuum behavior, the BNWF method based on the p–y curve has still been widely utilized because of its clear concept and less calculation cost.

At present, the API p–y curve for soft clay [8], which is most widely used in industry, is proposed based on the specific field test of small diameter piles. Its applicability to larger diameter pile foundations, such as large-diameter steel pipe piles for offshore wind turbines, has been questioned. P–y curves with better applicability urgently needs to be developed. This research has two main objectives: (a) clarify the defects of widely used API p–y curve for soft clays and (b) develop a new p–y curve of soft clay with better applicability for large diameter monopile foundations.

In order to achieve the above research purposes, the BNWF method based on the p–y curve and three-dimensional finite element method for analyzing the lateral responses of monopile are developed based on ABAQUS software; then, the accuracy of the proposed two methods are verified using the field test results of pile in soft clays [7]. Subsequently, the lateral responses of monopiles with different diameters are simulated employing the two methods. The defects of the API p–y curve are clarified by comparing the calculated results obtained by the two methods. Based on the above findings, a new p–y model for soft clay is proposed and the good applicability of the model to large diameter monopiles is validated by the above three-dimensional numerical simulation results. The novelty of this study is that a new p–y model with variable parameters was proposed, which is more versatile and offers a wider applicability to represent different soft clay and pile diameters.

Compared with the API p–y curve, the proposed p–y curve can better predict the lateral behavior of large diameter monopiles in soft clays.

2. Two Methods for Analyzing Lateral Responses of Piles in Soft Clays

2.1. Beam on Nonlinear Winkler Foundation (BNWF) Method

The BNWF model is developed based on ABAQUS software [39] to simulate the lateral monotonic load tests on teel pipe piles installed in soft clay at a site near the Sabine River [7]. The test pile had an outer diameter of 0.324 m with a wall thickness of 12.7 mm. The total length of the pile was 13.1 m with an embedded depth of 12.8 m. The lateral load was applied at 0.3 m above the ground surface. The average undrained shear strength of clays s_u = 14.4 kPa, effective unit gravity γ = 7.85 kN/m^3, elastic modulus of pile material E_P = 210 GPa, density ρ = 1820 kg/m^3, and Poisson's ratio v = 0.33 were all utilized.

The schematic diagram of the BNWF model is shown in Figure 1. The pile is discretized by 128 two-dimensional beam column elements (B21 element) and the length of each element L_e = 0.5 m. A nonlinear spring element is set at each element node to simulate the pile–soil interaction. The elastic modulus of the nonlinear spring is set according to the API p–y curve for soft clay. The spring resistance force $F = p \cdot L_e \cdot D$ (D is the outer diameter of the pile). The continuous variation of the elastic modulus of spring reflects the nonlinear relationship between the spring resistance force F (or lateral soil resistance) and the spring displacement y (the pile lateral relative displacement). The API p–y formula and relevant parameter values for this study are as follows:

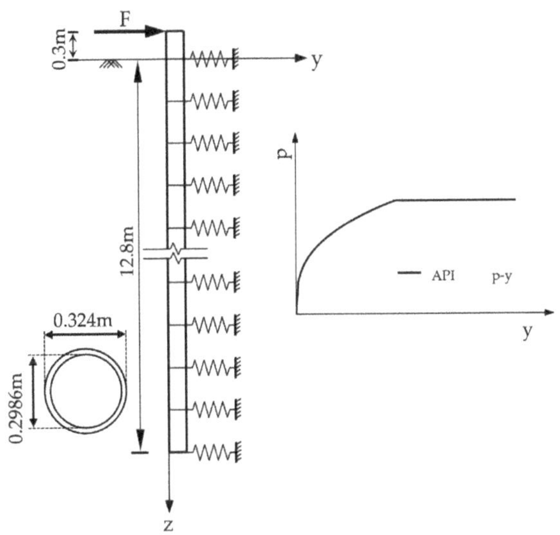

Figure 1. Schematic diagram of BNWF model for laterally loaded pile in soft clays.

$$\begin{cases} \frac{p}{p_u} = \left(\frac{y}{y_{50}}\right)^{1/3} & \frac{y}{y_{50}} \le 8 \\ \frac{p}{p_u} = 1 & \frac{y}{y_{50}} > 8 \end{cases} \quad (1)$$

where p is the mobilized lateral resistance (kPa); y is the local pile lateral displacement (m); y_{50} is the pile lateral displacement at half of ultimate soil resistance (m); and $y_{50} = 2.5\varepsilon_{50}D$, ε_{50} is the strain at one-half the maximum deviator stress in laboratory undrained compression tests of undisturbed soil samples. ε_{50} = 0.02 in this study. D is the pile diameter and D = 0.324 m. γ is effective unit weight of soil (kN/m^3) and γ = 7.85 kN/m^3. p_u is the ultimate resistance (kPa); $p_u = 3s_u + \gamma z + Js_u z/D$ for $z < z_r$ and $p_u = 9s_u$ for $z \ge z_r$. s_u is the undrained shear strength (kPa) and s_u = 14.4 kPa; z is the depth below the original seafloor (m); J is the dimensionless empirical constant with values ranging from 0.25 to 0.5 having been determined using field testing, and $J = 0.5$ in this study.

Figure 2a,b compare the calculated lateral deflection and bending moment profiles along the pile and those evaluated from the field measurements for four different pile head load levels. It can be seen that the calculated responses agree well with the test results for all load levels despite the slight deviations. One reason for the slight deviation is that the soil strength is idealized as uniform strength, but it is not uniform in the field test. Another reason is the limitation of the BNWF method itself, i.e., it ignores the continuity of soil, which will cause the error of calculation. This good agreement demonstrates that the developed BNWF model based on ABAQUS software can well predict the increase in lateral deflection and bending moment profiles with the increasing load level and the location of the maximum bending moment for different load levels. It also demonstrates the ability of the developed model to predict the lateral monotonic behavior of the pile in soft clays.

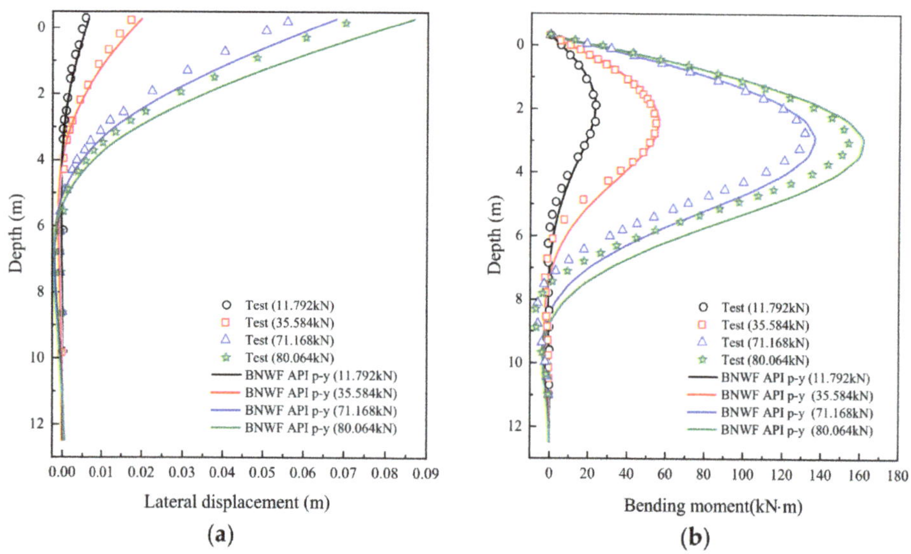

Figure 2. Comparison between calculated and measured test results based on the BNWF method. (**a**) Lateral deflection profile (**b**) Bending moment profile.

2.2. Three-Dimensional Finite Element Method (3D FEM)

A three-dimensional finite element model is established by employing the finite element software ABAQUS to simulate the interaction between soft clay and the pile foundation considering the specific conditions for Matlock's filed tests. Exploiting the symmetries of the geometry and loading conditions, the model simulates only one half of the system, as shown in Figure 3, to improve the computational efficiency. A sensitivity study indicates that a model of size $20\,D \times 50\,D$ (the cross section is a semicircle and D is the pile diameter) is sufficient to avoid boundary effects on the simulation results. The pile and the soils are all simulated using 5466 8-node linear brick elements with reduced integrations (C3D8R). Normal horizontal constraints are applied to the vertical boundaries and fixed constraints are applied to the bottom boundary. The top boundary is fully free. Lateral loads are applied to the pile according to the loading condition in the above field tests.

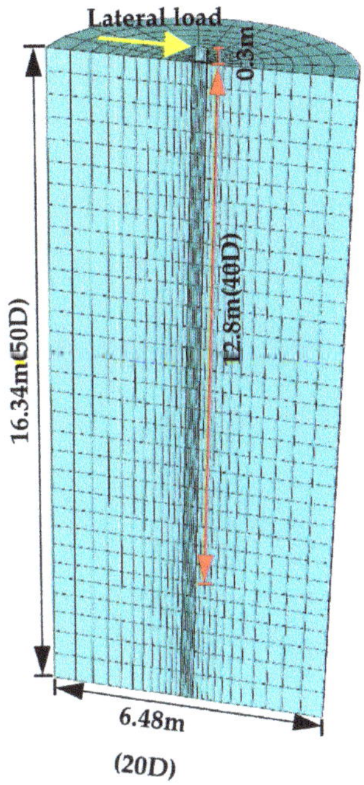

Figure 3. Three-dimensional finite element model for laterally loaded pile in soft clays.

The interaction between the monopile and surrounding soil is simulated utilizing the surface-to-surface contact method. The surface of the pile is defined as the master surface (relatively stiff) and the soil surface in contact with the pile is defined as the slave surface (relatively soft). Hard contact is set in the normal direction of the contact surface, which allows separation between the interface elements when they are subjected to tension. The penalty contact method is used in the tangential direction. When the two surfaces are in contact, the interface behavior is governed by Coulomb's friction theory. The critical friction shear stress τ_{crit} at the contact surface can be expressed by $\tau_{crit} = \mu \cdot p_c$ in terms of the frictional coefficient μ and contact pressure p_c, with μ = 0.3 in this study. When the shear stress at the contact surface exceeds τ_{crit}, the tangential slip occurs.

The behavior of clay is simulated by employing the elastic-perfectly plastic constitutive model with Mises yield criterion. The model parameters include the soil elastic modulus, Poisson's ratio, and shear strength, which can be determined using the laboratory tests or field tests. In this study, the model parameters are set according to the research of Matlock [7] and Hong et al. [37] as shown in Table 1. The material behavior of the hollow steel pipe pile is simulated using the linear elastic constitutive model. The model parameters are set according to the above steel properties. During the finite element simulation, geostatic stress balance analysis shall be carried out first, i.e., gravity is applied to the soil domain to establish the initial stress field. Meanwhile, the initial displacement field is set to 0. Then, lateral monotonic behavior of the pile is analyzed.

Table 1. Constitutive model parameters and relevant calculation parameters.

Parameter	Soil Elastic Modulus (Mpa)	Soil Poisson's Ratio	Soil Shear Strength (kPa)	Soil Effective Unit Weight (kN/m^3)	Pile–Soil Frictional Coefficient
Value	8	0.3	14.4	7.85	0.3

Figure 4 shows the comparison between the calculated results and test results of the lateral deflection and bending moment profiles along the pile for four different pile head load levels. The calculated results agree with the test results in the overall trend although there is some deviation. One reason for the deviation is that the soil strength is idealized as uniform strength, but it is not uniform in the field test. Another reason is that it is a simplification to use the elastic-perfectly plastic constitutive model to simulate the stress–strain responses of the soil; when more advanced constitutive models are adopted, better calculation results can be expected. This comparison demonstrates that the developed three-dimensional finite element model based on ABAQUS software can well predict the lateral monotonic behavior of the pile in soft clays.

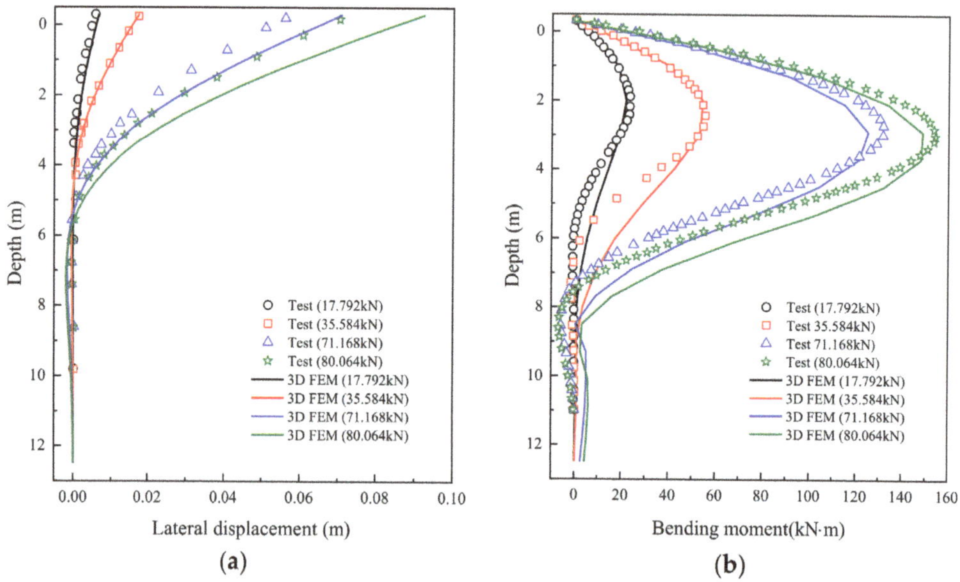

Figure 4. Comparison between calculated and measured test results based on 3D FEM. (**a**) Lateral deflection profile (**b**) Bending moment profile.

3. Lateral Behavior Analysis of Monopiles with Different Diameters

3.1. Numerical Model

Two methods for analyzing horizontally loaded piles, namely the BNWF method and 3D finite element method, have been established and validated, respectively, in Section 2 above. It should be noted that the BNWF model is actually a two-dimensional simplified calculation model of the pile–soil interaction; the accuracy of the calculation results employing this model mainly depends on the rationality of the p–y curve, hence it has certain limitations when using the currently questionable API p–y curve. In comparison, the three-dimensional finite element method has a better applicability for different pile diameters and site soil characteristics, although the calculation cost is relative larger. In this chapter, the above two methods will be used to simulate the lateral behavior of monopiles with different diameters. Using 3D finite element simulation results as a relatively correct reference standard, the limitation of the current API p–y curve is clarified by comparing the calculation results obtained from two methods.

Four diameter piles (1 m, 2 m, 3 m, 5 m) were selected to build both two-dimensional BNWF models and 3D finite element models. The total length of all piles is 60 m with an embedded depth of 50 m and the wall thickness is 30 mm. A lateral load of 2 MN is applied at the pile head (10 m above the mud surface). The three-dimensional finite element models of the four diameter piles are shown in Figure 5. In the 3D numerical model, the constitutive models and material parameters of pile and soil, pile–soil contact condition, boundary conditions, etc. are consistent with those in Section 2 above, which will not be repeated here due to space limitations. The setting of the two-dimensional model is also consistent with those in Section 2 except for the different pile sizes.

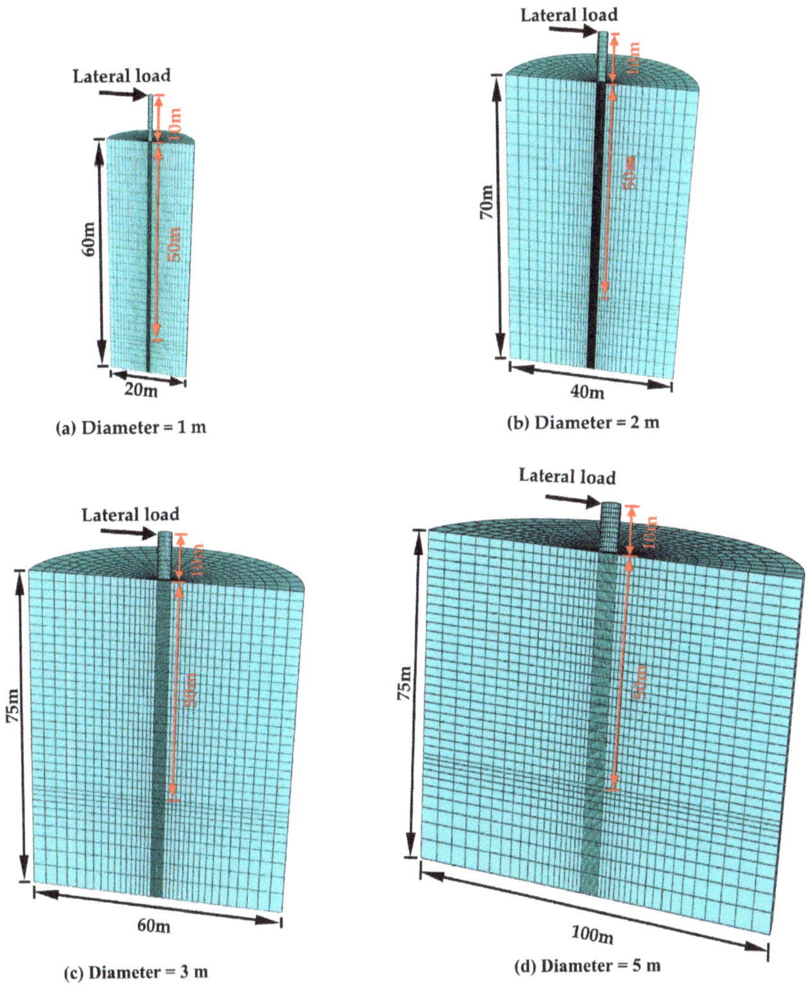

Figure 5. 3D finite element model for laterally loaded pile with different diameter in soft clays.

3.2. Comparison of Predicted Lateral Behavior for Two Methods

Figure 6 compares the lateral response of piles with various diameters obtained using the BNWF method based on the API p–y curve and 3D FEM when the pile head load equals 250 kN. It can be seen from Figure 6a that the nonlinearity of pile head load–displacement curve becomes more significant as the pile diameter decreases. The displacement of the pile head is greater for the smaller pile diameter under the same lateral load. The prediction result of displacement based on the BNWF method is larger than that of the 3D FEM, which indicates that the BNWF method underestimates the bearing capacity of piles. As shown in Figure 6b, the lateral deflection of the pile along the depth (range from −10 to 10) increases significantly as the pile diameter decreases, which is because the pile becomes more flexible due to the decrease in the stiffness. The lateral deflection profile obtained by the BNWF method is larger than 3D FEM and the difference between the two calculation results is more significant for the smaller diameter piles. It can be observed from Figure 6c that the bending moment of the pile along the depth increases significantly as the pile diameter increases and the location of the maximum bending moment moves downward gradually. The bending moment profile obtained by the BNWF method is larger than 3D FEM and the difference is more significant for the larger diameter piles. From the above comparison, it can be concluded that the BNWF method based on the API p–y curve will overestimate the deflection and bending moment and underestimate the ultimate bearing capacity of the large diameter pile.

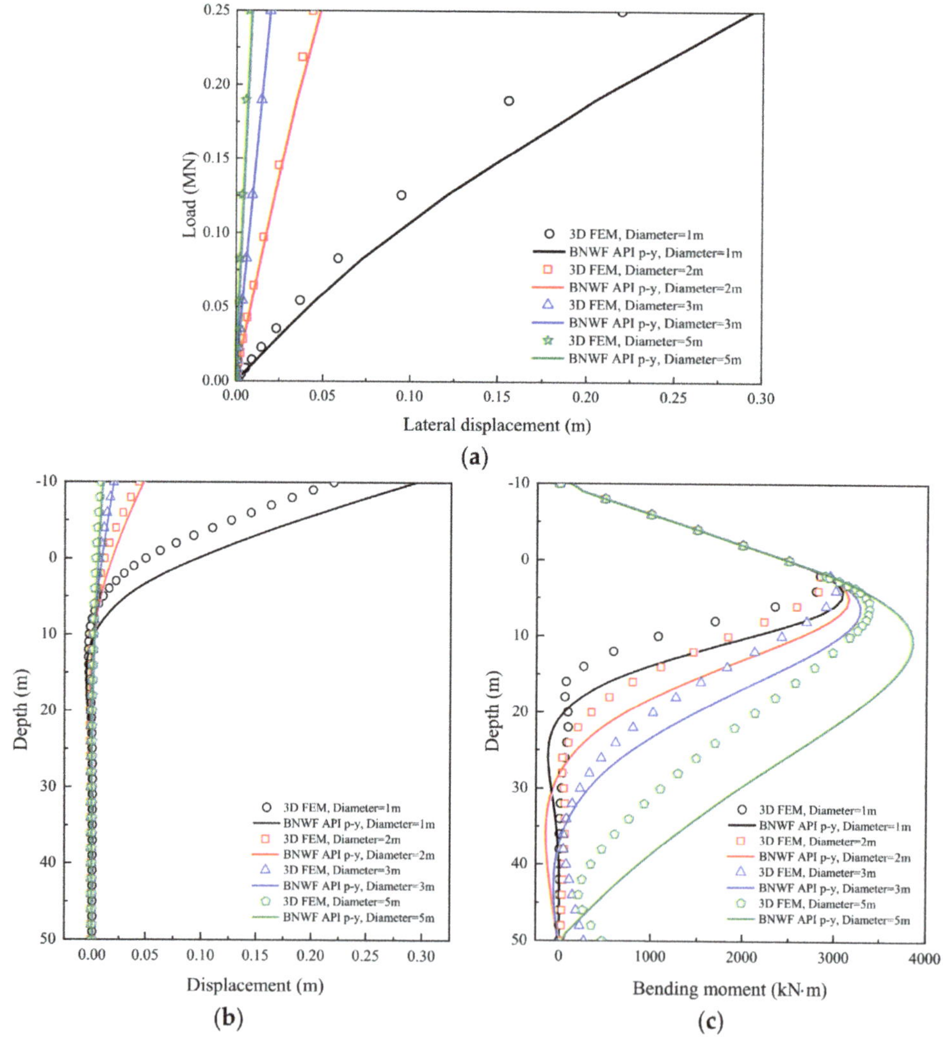

Figure 6. Comparison of lateral response of piles obtained by BNWF method (API p–y curve) and 3D FEM. (**a**) Load–displacement curve of pile head. (**b**) Lateral deflection profile. (**c**) Bending moment profile.

3.3. Limitations of API p–y Curve for Soft Clay

The pile–soil interaction is simulated through the surface–surface contact mode for the 3D finite element model. The soil resistance value P at a certain depth of the pile shaft is calculated by extracting the contact force on the contact surface. Figure 7a shows the typical distribution of the calculated contact normal force (CNF) and the contact shear force (CSF) around the pile. Figure 7b shows the diagrammatic sketch for the x-direction component of the CNF (CNF1) and the x-direction component of the CSF (CSF1) at a certain position on the pile cross-section. Figure 7b,c shows the typical distribution of CNF1 and CSF1 around the pile. It can be observed that CNF1 is the largest in the x-axis direction and the smallest at the edges on both sides of the section (almost 0); on the contrary, CSF1 is the smallest (almost 0) in the x-axis direction and the largest at the edges on both sides of the section.

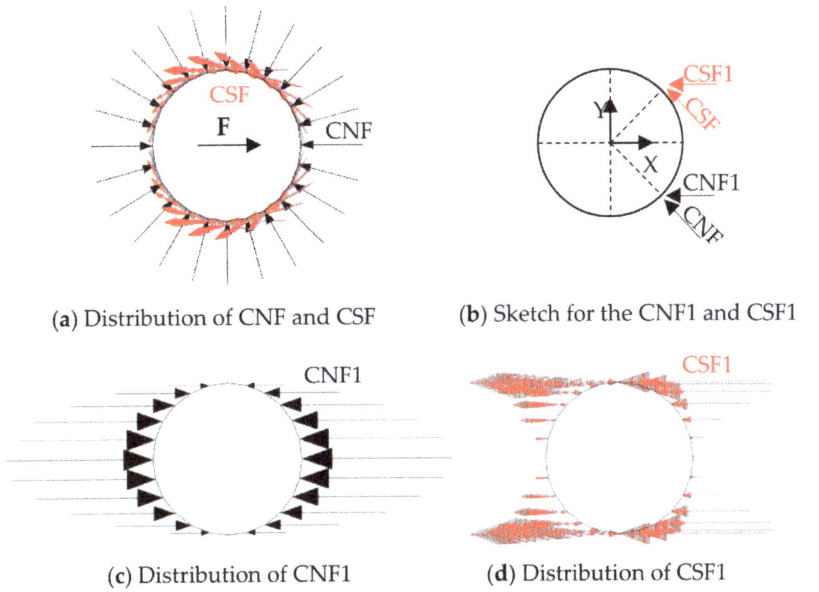

Figure 7. Typical distribution of the calculated contact normal force (CNF) and contact shear force (CSF).

The soil resistance P at a certain depth of the pile shaft can be determined by the sum of CNF1 and CSF1 around the pile, as represented by:

$$P = \sum_{i=1}^{n} (CNF_1 + CSF_1) \quad (2)$$

where n is the number of nodes outside the pile cross-section at a certain depth of the pile shaft. Then, the p–y curve can be determined based on the numerical simulation results.

Figure 8 compares the API p–y curve with the p–y curve obtained using the 3D finite element method. These p–y curves correspond to different depths (depth = 1 m, 3 m, 5 m, and 10 m) for various pile diameters (diameter = 1 m, 2 m, 3 m, and 5 m). For these larger diameter piles, there are significant difference between the API p–y curve and the p–y curve obtained using the 3D finite element method. The comparison demonstrates: (1) the API p–y curve significantly underestimates the ultimate soil reaction of soft clays and the degree of the underestimation increases as the soil depth increases; (2) the API p–y curve significantly underestimates the initial stiffness of soft clays. This conclusion can well explain why the BNWF method based on the API p–y curve overestimates the deflection and bending moment while it underestimates the ultimate bearing capacity of the large diameter pile.

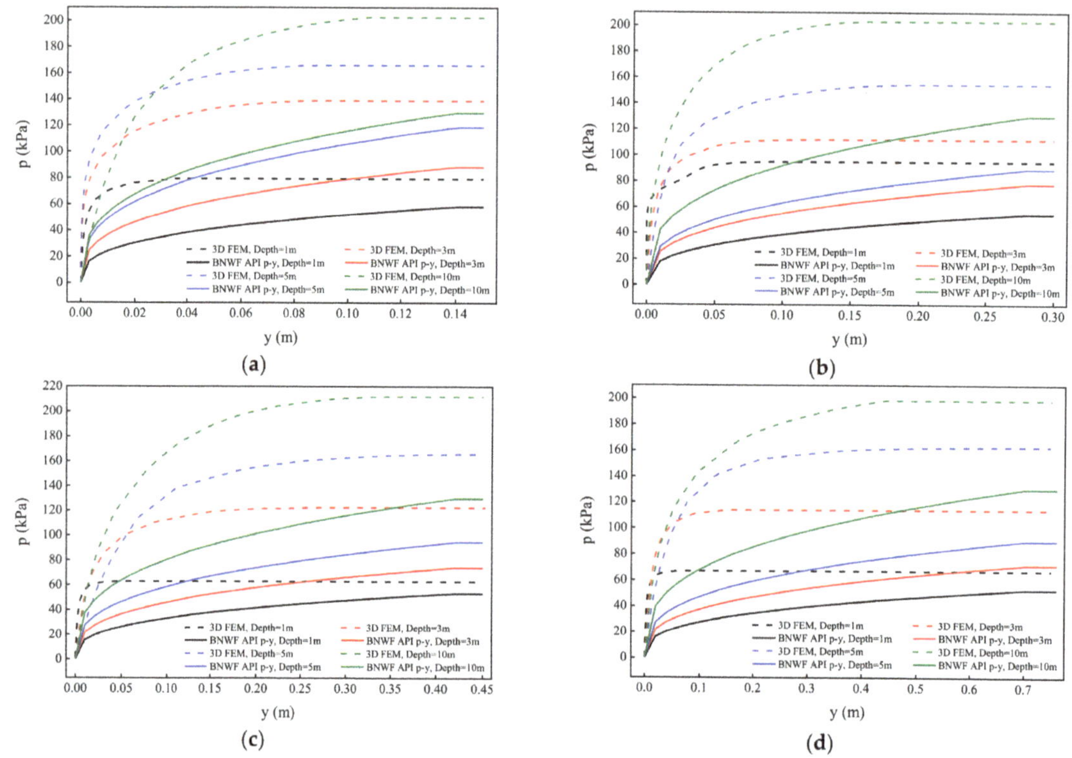

Figure 8. Comparison of the API p–y curve and the p–y curve obtained by 3D finite element method. (**a**) Diameter = 1 m, (**b**) Diameter = 2 m, (**c**) Diameter = 3 m, and (**d**) Diameter = 5 m.

4. New p–y Curve for Soft Clay

4.1. New p–y Formula and Parameter Influence

In order to improve the shortcomings of the API p–y curve, a new p–y curve is proposed in this section, as shown in the formula:

$$p = \frac{B \cdot G_{\max} \cdot y}{A \cdot D}\left(1 - \frac{y}{\frac{D \cdot p_{\max}}{B \cdot G_{\max}} + y}\right)^{A} \quad (3)$$

where $G_{\max}$ is the maximum shear modulus and $G_{\max}$ is usually considered to be a multiple of s_u for soft clay according to the engineering experience (e.g., typical range of $G_{\max}$ = from 300 s_u to 1800 s_u); parameters A and B control the shape of the p–y curve. $p_{\max}$ is the ultimate lateral soil resistance, which is calculated based on the expression proposed by Murff and Hamilton [40] for shear strength profiles approximately linearly increasing with depth, i.e.,:

$$\begin{aligned} &p_{\max} = N_p \cdot s_u \\ &N_p = 12 - 4 \cdot \exp\left(\frac{-\zeta \cdot z}{D}\right) \\ &\zeta = 0.25 + 0.05 \cdot \lambda \; for \; \lambda < 6 \\ &\zeta = 0.55 \qquad\qquad for \; \lambda \geq 6 \\ &\lambda = s_{u0}/(s_{u1} \cdot D) \end{aligned} \quad (4)$$

where s_u is the undrained shear strength of soil at the point in question; N_p is a nondimensional lateral bearing factor; γ' is the soil submerged unit weight; z is the soil depth below the original seafloor; D is the pile outside the diameter; s_{u0} is the shear strength intercept at the seafloor; and s_{u1} is the rate of increase in shear strength with depth.

In order to investigate the influence of the relevant parameters on the p–y curve, the soil properties are assumed as follows: s_u = 14.4 kPa (s_{u0} = 14.4 kPa, s_{u1} = 0 in Equation (4)); G_{max} = 500 s_u = 7.2 MPa; pile diameter D = 3 m; p_{max} = 170 kPa according to Equation (4) when soil depth z = 15 m.

A series of A values are set to investigate its influence on the shape of the p–y curve (B = 1 for all cases). Figure 9a,b show the p–y curves and the normalized p–y curves (p is normalized by p_{max} and y is normalized by D) for various A values. It can be observed that the decrease in the soil stiffness (tangent modulus) becomes slower as the displacement increases and the variation of the p–y from linear to nonlinear becomes more gradual and starts earlier as A increases. It is found that the p–y (or normalized p–y) curve has insignificant change when A is less than 0.01 and hence A usually ranges from 0.1 to 10 to fit the actual p–y curves. It is worth noting that the current model can be simplified into Kondner's hyperbolic p–y model [9] when A = 1, as represented by:

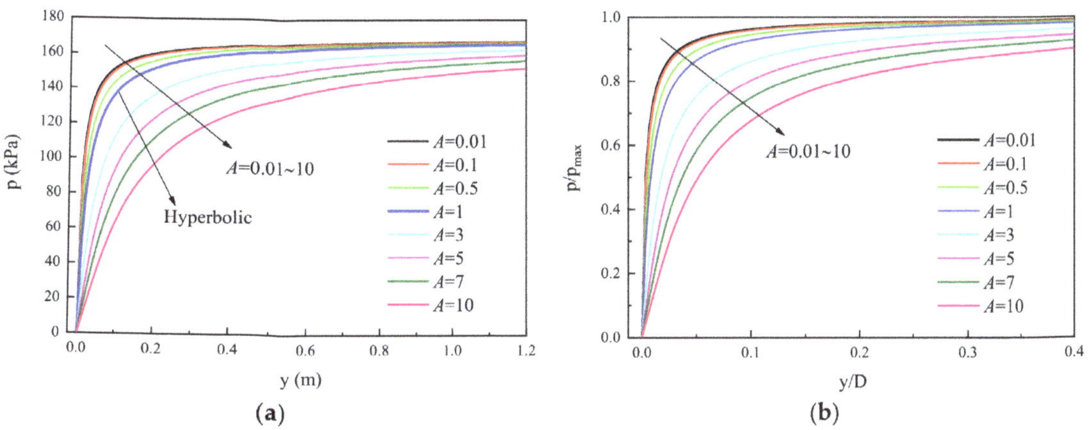

Figure 9. Effect of parameter A on the proposed p–y curve. (**a**) p–y curve (**b**) normalized p–y curve.

$$p = \frac{y}{\frac{1}{\frac{B \cdot G_{max}}{D}} + \frac{y}{p_{max}}} \quad (5)$$

For this case, the initial maximum tangent modulus of the p–y curve is $B \cdot G_{max}/D$.

A series of B values are set to investigate its influence on the shape of the p–y curve (A = 1 for all cases). Figure 10a,b show the p–y curves and the normalized p–y curves for various B values. Contrary to the effect of parameter A on the p–y curve, the decrease in the soil stiffness (tangent modulus) becomes faster as the displacement increases, and the variation of p–y from linear to nonlinear becomes more abrupt as B increases. When parameter B exceeds 10, it has little effect on the p–y (or normalized p–y) curve and hence B usually ranges from 1 to 10 to fit the actual p–y curves. Obviously, the proposed p–y model with variable A and B is more versatile and offers wider applicability to represent different soft clays and pile diameters.

The lateral responses of the piles with various diameters are predicted employing the BNWF method based on the proposed API p–y curve. The prediction results are compared with the 3D finite element results on Section 3.2, as shown in Figure 14. The prediction results including the load–displacement curve of the pile head, lateral deflection profile, and bending moment profile agree well with the 3D finite element results. This comparison demonstrates that, compared with API p–y curve, the proposed p–y curve can better predict the lateral behavior of piles in soft clays and has better applicability to large-diameter piles.

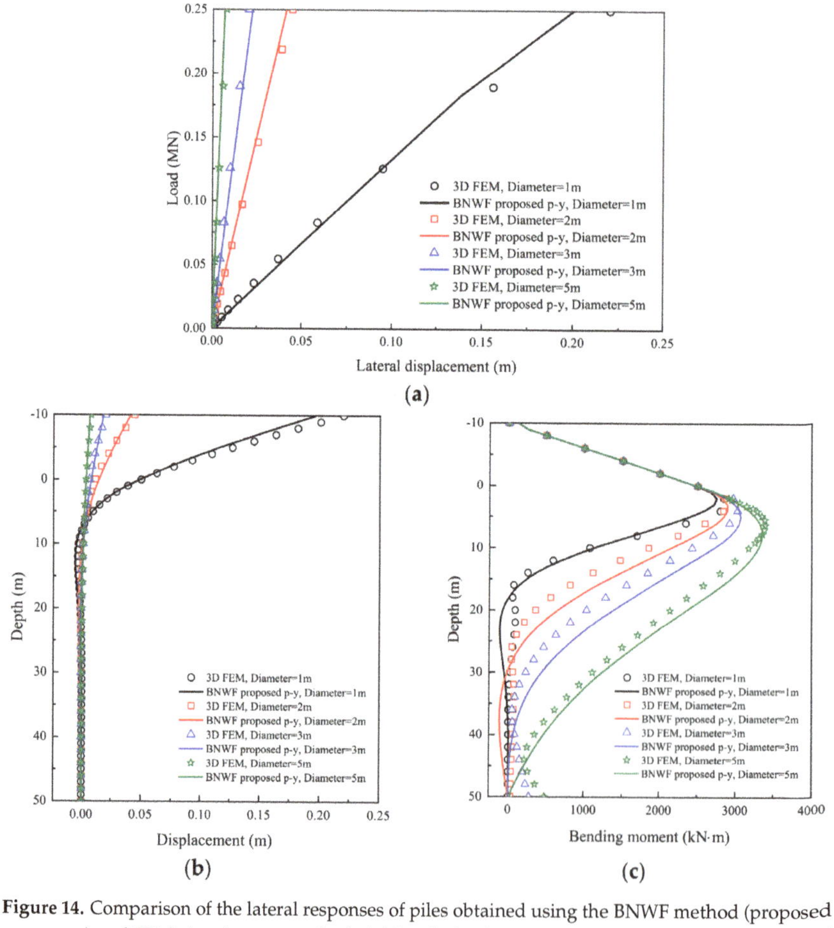

Figure 14. Comparison of the lateral responses of piles obtained using the BNWF method (proposed p–y curve) and 3D finite element method. (**a**) Load–displacement curve of pile head. (**b**) Lateral deflection profile (**c**) Bending moment profile.

5. Summary and Conclusions

5.1. Main Conclusions

(1) The BNWF method based on the p–y curve and 3D FEM for analyzing the lateral responses of monopiles are developed based on ABAQUS software and the accuracy of the two methods is validated using the field test results of the piles in soft clays.

(2) The lateral responses of monopiles with different diameters are simulated by employing the two methods. It is found that the BNWF method based on the API p–y curve overestimates the deflection and bending moment while it underestimates the ultimate bearing capacity of the large diameter pile.

(3) The API p–y curve significantly underestimates the ultimate soil reaction of soft clays and the degree of the underestimation increases as the soil depth increases; it also significantly underestimates the initial stiffness of the soft clays. These findings are supported by previous research results [12,15].
(4) A new p–y model with variable parameters A and B was proposed that is more versatile and offers wider applicability to represent different soft clay and pile diameters. The current model can be simplified into the hyperbolic p–y model when A =1.

5.2. Significance and Application

As the most popular foundation type for supporting offshore wind turbines, large diameter monopile foundations must be designed for lateral loads such as winds, waves, and currents. The Beam on Nonlinear Winkler Foundation (BNWF) method has been widely used because of its clear concept and lower calculation cost. Compared with the three-dimensional finite element method, the BNWF method is easier to implement numerically. This method can be realized using ABAQUS software or, alternatively, by using other special calculation software for pile foundation such as LPILE or by compiling simple MATLAB programs. Hence, it has better applicability in engineering practice. The selection of a reasonable p–y curve is critical to the calculation accuracy of this method. This paper proposes a new p–y curve with better versatility and applicability. Compared with the API p–y curve, the proposed p–y curve can better predict the lateral behavior of piles in soft clays, such as the load–displacement curve of the pile head, lateral deflection profile, and bending moment profile, and has better applicability to large-diameter piles. However, it should be noted that the proposed p–y curve is mainly applicable to soft clay and not to other types of soil such as stiff clays and sands. When applied to engineering practice, the proposed p–y model can play a good role in the design of monopile foundations supporting offshore wind turbines in soft clays.

Author Contributions: Data curation, Q.L. and J.L.; Funding acquisition, M.W. (Mingyuan Wang); Investigation, M.W. (Mingyuan Wang) and J.L.; Methodology, M.W. (Mingyuan Wang), M.W. (Miao Wang) and X.C.; Resources, Q.L.; Software, M.W. (Miao Wang) and X.C.; Supervision, Q.L.; Validation, M.W. (Miao Wang); Writing—original draft, X.C. All authors have read and agreed to the published version of the manuscript.

Funding: This work was supported by the National Natural Science Foundation of China (52108334), the Natural Science Foundation of Tianjin City (20JCYBJC00570), and the Science and Technology Projects of Powerchina Huadong Engineering Corporation Limited (KY2020-KC-08-01-2021).

Data Availability Statement: The data presented in this study are available on request from the corresponding author. The data are not publicly available due to the follow-up research.

Conflicts of Interest: The authors declare no conflict of interest.

References

1. Carter, J.M.F. North Hoyle offshore wind farm: Design and build. *Proc. Inst. Civ. Eng.-Energy* **2007**, *160*, 21–29. [CrossRef]
2. Oh, K.Y.; Nam, W.; Ryu, M.S. A review of foundations of offshore wind energy convertors: Current status and future perspectives. *J. Renew. Sustain. Energy Rev.* **2018**, *88*, 16–36. [CrossRef]
3. Wu, X.; Hu, Y.; Li, Y. Foundations of offshore wind turbines: A review. *J. Renew. Sustain. Energy Rev.* **2019**, *104*, 379–393. [CrossRef]
4. Hamilton, B.; Battenberg, L.; Bielecki, M.; Bloch, C.; Decker, T.; Frantzis, L.; Paidipati, J.; Wickless, A.; Zhao, F. *Offshore Wind Market and Economic Analysis: Annual Market Assessment*; Navigant Consulting, Inc.: Burlington, MA, USA, 2013.
5. Esteban, M.D.; Lopez-Gutierrez, J.S.; Negro, V.; Matutano, C.; Garcia-Flores, F.M.; Millan, M.A. Offshore wind foundation design: Some key issues. *J. Energy. Resour. Technol.* **2015**, *137*, 051211. [CrossRef]
6. McClelland, B.; Focht, J.A. Soil modulus for laterally loaded piles. *Trans. ASCE* **1956**, *182*, 1–22. [CrossRef]
7. Matlock, H. Correlations for design of laterally-loaded piles in soft clay. In Proceedings of the 2nd Offshore Technology Conference, Houston, TX, USA, 22–24 April 1970; p. 1204.
8. API Recommended Practice 2GEO. *Geotechnical and Foundation Design Considerations*, 1st ed.; Addendum 1; American Petroleum Institute: Washington, DC, USA, 2014.
9. Kondner, R.L. Hyperbolic stress-strain response: Cohesive soils. *J. Soil. Mech. Found. Div.—ASCE* **1963**, *89*, 115–143. [CrossRef]
10. Georgiadis, M.; Anagnostopoulos, C.; Saflekou, S. Cyclic lateral loading of piles in soft clay. *Geotech. Eng.* **1992**, *23*, 47–60.

11. Dewaikar, D.M.; Patil, P.A. A new hyperbolic *p-y* curve for laterally loaded piles in soft caly. *Found. Anal. Des.* **2006**, 152–158.
12. Jeanjean, P. Re-assessment of *p-y* curves for soft clays from centrifuge testing and finite element modelling. In Proceedings of the Offshore Technology Conference (OTC 20158), Houston, TX, USA, 4–7 May 2009.
13. Klinkvort, R.T.; Hededal, O. Effect of load eccentricity and stress level on monopile support for offshore wind turbines. *Can. Geotech. J.* **2014**, *51*, 966–974. [CrossRef]
14. Rathod, D.; Muthukkumaran, K.; Sitharam, T.G. Effect of slope on *p-y* curves for laterally loaded piles in soft clay. *Geotech. Geol. Eng.* **2018**, *36*, 1509–1524. [CrossRef]
15. Zhu, B.; Zhu, Z.J.; Li, T.; Liu, J.C.; Liu, Y.F. Field tests of offshore driven piles subjected to lateral monotonic and cyclic loads in soft clay. *J. Waterw. Port. C ASCE* **2017**, *143*, 05017003. [CrossRef]
16. Xu, D.S.; Xu, X.Y.; Li, W. Field experiments on laterally loaded piles for an offshore wind farm. *Mar. Struct.* **2020**, *69*, 102684. [CrossRef]
17. Wang, Z.; Xie, X.; Wang, J. A new nonlinear method for vertical settlement prediction of a single pile and pile groups in layered soils. *Comput. Geotech.* **2012**, *45*, 118–126. [CrossRef]
18. Zhang, C.; Yu, J.; Huang, M. Winkler load-transfer analysis for laterally loaded piles. *Can. Geotech. J.* **2016**, *53*, 1110–1124. [CrossRef]
19. Choo, Y.W.; Kim, D. Experimental development of the *p-y* relationship for large-diameter offshore monopiles in sands: Centrifuge tests. *J. Geotech. Geoenviron. Eng.* **2016**, *142*, 04015058. [CrossRef]
20. Boonyatee, T.; Lai, Q.V. A non-linear load transfer method for determining the settlement of piles under vertical loading. *Int. J. Geotech. Eng.* **2017**, *14*, 206–217. [CrossRef]
21. Zhang, Y.; Andersen, K.H. Scaling of lateral pile *p-y* response in clay from laboratory stress-strain curves. *Mar. Struct.* **2017**, *53*, 124–135. [CrossRef]
22. Elkasabgy, M.; El Naggar, M.H. Lateral performance and p–y curves for large-capacity helical piles installed in clayey glacial deposit. *J. Geotech. Geoenviron. Eng.* **2019**, *145*, 04019078. [CrossRef]
23. Li, S.; Yu, J.; Huang, M.; Leung, C.F. Application of T-EMSD based *p-y* curves in the three-dimensional analysis of laterally loaded pile in undrained clay. *Ocean. Eng.* **2020**, *206*, 107256. [CrossRef]
24. Zhang, Y.; Andersen, K.H. Soil reaction curves for monopiles in clay. *Mar. Struct.* **2019**, *65*, 94–113. [CrossRef]
25. El Naggar, M.H.; Novak, M. Nonlinear analysis for dynamic lateral pile response. *Soil Dyn. Earthq. Eng.* **1996**, *15*, 233–244. [CrossRef]
26. Gerolymos, N.; Gazetas, G. Phenomenological model applied to inelastic response of soil-pile interaction systems. *Soils Found.* **2005**, *45*, 119–132. [CrossRef]
27. Allotey, N.; El Naggar, M.H. Generalized dynamic Winkler model for nonlinear soil-structure interaction analysis. *Can. Geotech. J.* **2008**, *45*, 560–573. [CrossRef]
28. Heidari, M.; El Naggar, M.H.; Jahanandish, M.; Ghahramani, A. Generalized cyclic *p-y* curve modeling for analysis of laterally loaded piles. *Soil Dyn. Earthq. Eng.* **2014**, *63*, 138–149. [CrossRef]
29. Liang, F.Y.; Chen, H.B.; Jia, Y.J. Quasi-static *p-y* hysteresis loop for cyclic lateral response of pile foundations in offshore platforms. *Ocean Eng.* **2018**, *148*, 62–74. [CrossRef]
30. Wu, K.; Chen, R.; Li, S. Finite element modeling of horizontally loaded monopile foundation of large scale offshore wind turbine in non-homogeneity clay. *IEEE Softw.* **2009**, *2*, 329–333.
31. Abdel-Rahman, K.; Achmus, M. Numerical modeling of the combined axial and lateral loading of vertical piles. In Proceedings of the 6th European Conference on Numerical Methods in Geotechnical Engineering, Graz, Austria, 6–8 September 2006; pp. 575–581.
32. Haiderali, A.; Madabhushi, G. Three-dimensional finite element modelling of monopiles for offshore wind turbines. In Proceedings of the World Congress on Advances in Civil, Environmental and Materials Research, Seoul, Korea, 26–29 August 2012; pp. 3277–3295.
33. Haiderali, A.; Cilingir, U.; Madabhushi, G. Lateral and axial capacity of monopiles for offshore wind turbines. *Indian Geotech. J.* **2013**, *43*, 181–194. [CrossRef]
34. Cheng, X.L.; Wang, T.J.; Zhang, J.X.; Liu, Z.X.; Cheng, W.L. Finite element analysis of cyclic lateral responses for large diameter monopiles in clays under different loading patterns. *Comput. Geotech.* **2021**, *134*, 104104. [CrossRef]
35. Achmus, M.; Abdel-Rahman, K.; Kuo, Y.S. Numerical modelling of large diameter steel piles under monotonic and cyclic horizontal loading. In Proceedings of the Tenth International Symposium on Numerical Models in Geomechanics, Rhodes, Greece, 25–27 April 2007; Taylor Francis: London, UK, 2007; pp. 453–459.
36. Heidari, M.; Jahanandish, M.; El Naggar, H.; Ghahramani, A. Nonlinear cyclic behavior of laterally loaded pile in cohesive soil. *Can. Geotech. J.* **2014**, *51*, 129–143. [CrossRef]
37. Hong, Y.; He, B.; Wang, L.Z.; Wang, Z.; Ng, W.W.C.; Masin, D. Cyclic lateral response and failure mechanisms of semi-rigid pile in soft clay: Centrifuge tests and numerical modelling. *Can. Geotech. J.* **2017**, *54*, 806–824. [CrossRef]
38. He, B.; Lai, Y.; Wang, L.; Hong, Y.; Zhu, R. Scour effects on the lateral behavior of a large-diameter monopile in soft clay: Role of stress history. *J. Mar. Sci. Eng.* **2019**, *7*, 170. [CrossRef]
39. Dassault Systemes Simulia Corp. *Abaqus Analysis User's Manual Version 6.14*; Simulia: Johnston, RI, USA, 2014.
40. Murff, J.D.; Hamilton, J.M. P-ultimate for undrained analysis of laterally loaded piles. *J. Geotech. Eng.* **1993**, *119*, 91–107. [CrossRef]

sustainability MDPI

Article

Calculation Method for Uplift Capacity of Suction Caisson in Sand Considering Different Drainage Conditions

Chenggen Xu [1], Haitao Jiang [1], Mengtao Xu [2,*], Decheng Sun [1] and Shengjie Rui [2,3]

[1] CGN Rudong Offshore Wind Power Co., Ltd., Rudong 226400, China
[2] Key Laboratory of Offshore Geotechnics and Material of Zhejiang Province, College of Civil Engineering and Architecture, Zhejiang University, Hangzhou 310058, China
[3] Norwegian Geotechnical Institute, 0484 Oslo, Norway
* Correspondence: xmt736@zju.edu.cn

Abstract: Uplift capacity of suction caissons is one of the main concerns in the design of jackets with multi-caissons supported offshore wind turbine. The uplift movement of suction caissons leads to soil stress variation and increases the difficulty to predict the uplift capacity. In this paper, a calculation method considering soil stress release and differential pressure contribution is proposed to predict the uplift capacity of caisson. Firstly, a series of numerical simulations based on the SANISAND model are conducted to study the uplift responses of suction caisson in sand, and it is verified with centrifuge test results. Considering the soil drainage condition during caisson being pulled out, the fully drained, partially drained and undrained are divided, and an equation is provided to assess differential pressure beneath the caisson lid incorporating the effects of main factors. Based on the above simulation results, a calculation method is proposed to calculate the uplift capacity of caissons. The prediction results are compared with the centrifuge model tests and previous studies, which indicate that the prediction accuracy is much improved. This proposed method contributes to the more accurate assessment of uplift capacity of suction caisson in sand.

Keywords: suction caisson; uplift capacity; calculation method; drainage condition; differential pressure

1. Introduction

In recent years, offshore wind energy has developed rapidly due to the characteristics of high wind speed, low turbulence and no occupation of farmland [1]. A variety of offshore wind turbine (OWT) foundations have emerged, e.g., gravity, monopile, suction caisson, jacket foundation, floating foundation, etc. For water depth of more than 30 m, the jacket foundation is always recommended [2–4]. The jacket with multi-caissons has great conveniences in transportation and installation, becoming a popular foundation form for OWTs [5].

During service, a large overturning moment generated by environment loads is applied to the jacket with multi-caissons supported OWT, and it is mainly transmitted to every caisson in the form of axial load [6,7]. The OWTs may be suddenly subjected to extreme environmental loads, such as a storm, resulting in transient tensile loading on upwind caisson, causing it at the risk of being pulled out.

The uplift capacity of suction caisson consist of differential pressure beneath the caisson lid and skirt wall friction [8–10]. The obvious permeability of sand results in a significant seepage field in soil domain [11], which has a prominent influence on the development of differential pressure. Pore pressure was a simulated transmission in numerical simulation by adding a thin layer of "water elements" between caisson lid and soil [12–17]. Vaitkunaite [18] conducted a series of large-scale model tests to study the development of differential pressure under axial loading. Hong [19] studied the influence of pullout velocity on drainage conditions through model experiments, which neglected the scale effect so that the impact of seepage cannot be amplified into the prototype.

The seepage field also has a significant impact on the soil stress state, causing the change in uplift capacity of suction caisson. Considering the soil stress reduction [20], Houlsby [21] proposed an analytical solution to calculate the tensile capacity of suction caisson under rapid loading, and the cavitation effect and soil liquefaction were considered. However, the calculation accuracy depends on the determination of stress reduction region, which brings difficulty to the engineering application. Senders [22] conducted a series of centrifuge tests to investigate the development of differential pressure and skirt wall friction during caisson being pulled out, and proposed a spring—damper model to predict the differential pressure beneath the lid and skirt wall friction. Vicent [23,24] performed model tests to investigate the impacts of loading rate and embedment depth on the pullout capacity of caisson foundations in sand.

The design code DNVGL [25] pointed out that the design of suction caisson will be conservative when the drainage mechanisms cannot be determined with good confidence. In the codes API [26] and OWA [27], the capacity of suction caisson is designed separately aiming at fully drained and undrained two conditions. The soil exhibits various failure mechanisms under different drainage conditions during caisson being pulled out [18,28]. However, these previous studies and design codes have not clearly classified the drainage conditions in sand and no suitable calculation scheme of differential pressure is provided.

In this paper, the SANISAND model is adopted to study the uplift responses of suction caisson considering the influence of stress state on soil mechanical responses. The effects of soil permeability, pullout velocity, aspect ratio, lateral earth pressure coefficient and soil-structure interface friction angle on the instantaneous pullout responses of caisson are studied. Then, the various drainage conditions are divided and an equation is proposed to evaluate the differential pressure during caisson pullout. Under different pullout velocity, the variation range of uplifted capacity and soil stress state are investigated. Then, the calculation method of uplift capacity is proposed combining the above simulation results. Compared with the centrifuge tests and preceding studies, the accuracy of the proposed method is validated.

2. Numerical Model and Its Verification

2.1. Finite Element Model

An axisymmetric model was built to research the pullout characteristics of suction caisson in sand using ABAQUS [29]. It is assumed that the caisson was "wish-in-place", ignoring its installation process. As shown in Figure 1, the finite element model consists of caisson, homogeneous saturated sand and a thin layer of soft "water elements". The tensile loading was applied by vertical displacement and its action point was caisson centre.

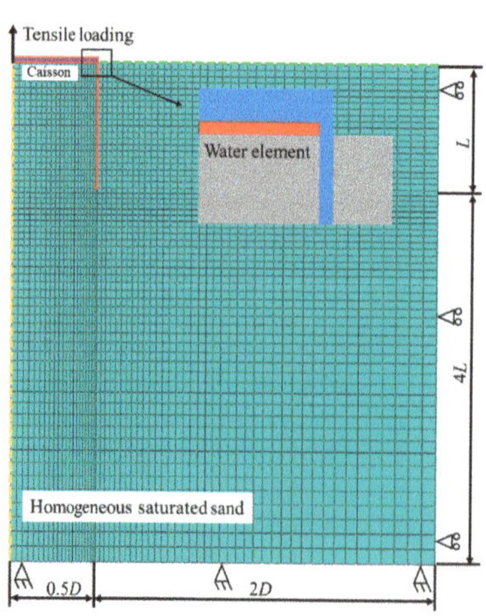

Figure 1. Finite element model.

2.1.1. Mesh Element

The mesh element CAX4 was used to simulate the caisson, and the element CAX4P containing pore pressure calculation was used for homogeneous saturated sand. The mesh width near the skirt reaches 0.05 m, which is fine enough verified by the mesh sensitivity study.

A thin layer of soft "water elements" was adopted to simulate the trapped water between the caisson lid and internal soil plug [12,13,30]. As normal coupled pore fluid elements, the water elements were defined as isotropic liner elastic property with high permeability and elasticity. The permeability coefficient was set to a high value of 1 m/s, and the Young's modulus and Poisson's ratio were 10^{-5} kPa and 0, respectively. In this way, the water elements can transfer pore pressure without transmitting effective stress.

2.1.2. Interface Properties and Boundary Conditions

The water elements were tied to the caisson lid and soil plug to maintain displacement coordination. The interfaces between skirt wall and soil were simulated using the master-slave contacts. The ultimate shear stress is calculated by the Coulomb criterion and can be expressed by $\mu\sigma'_h$, where μ is the friction coefficient of soil-structure and σ'_h is the horizontal soil stress on caisson. The elastic sliding distance of skirt-soil interface was set to 0.5 mm [15]. Ignoring the tangential behavior between caisson tip and soil, the separation was allowed.

In Figure 1, the left boundary is the symmetry axis of this model, and the right boundary is only constrained to lateral displacement and free for drainage. The bottom boundary is fixed, and the top boundary is free surface with zero excess pore pressure. The radius and the depth of soil domain are 2.5D and 5L, respectively, which can eliminate the boundary effects [14].

2.1.3. Soil Constitutive Model

The boundary surface soil model proposed by Dafalias [31,32], which is developed from the concept of critical state soil mechanics and belongs to one kind of SANISAND models [33], is chosen in this paper. The soil constitutive model has been certified to be able to simulate the sand stress-strain relationships under unidirectional and cyclic loads. This model is applicable for a range of relative density, from loose to medium dense, under different drainage conditions and consolidation stresses. Based on the framework of critical state soil mechanics, the bounding surface and phase transformation surface (determining the sand dilates or contracts) are imported.

2.2. Model Verification

Senders [22] performed a series of centrifuge tests to investigate the uplift characteristics of suction caisson at 100 g. For the model caisson, its outer diameter, skirt wall thickness and length are 60 mm, 0.4 mm and 60 mm, respectively. The interface friction angle of soil structure δ is 22°. The seabed is the silica sand with relative density D_r = 60%, mass median diameter D_{50} = 0.2 mm, initial void ratio e = 0.7357 and the internal friction angle φ = 43°. Silicon oil was used as the pore fluid and the permeability coefficient k is 2.3×10^{-4} m/s. The buoyant unit weight γ' and specific weight of water γ_w are both 10 kN/m^3. According to the results of centrifuge tests, Senders [22] calculated the lateral earth pressure coefficient, which is equal to 0.8. For the silica sand used in centrifuge tests of Senders [22], Tran [34] conducted a series of laboratory tests to measure its basic parameters. Based on the laboratory tests of Tran [34], Shen [15] calibrated the SANISAND model parameters of silica sand.

The centrifuge tests are divided into drained uplift (the air hole of caisson lid is opened) with the pullout velocity of 0.1 mm/s and partially drained uplift (the air hole of caisson lid is closed) with the pullout velocity of 0.1 mm/s and 0.5 mm/s. Figure 2 shows the comparison of simulation results with centrifuge tests. For the differential pressure (in Figure 2a) of skirt wall friction (in Figure 2b) and total vertical force (in Figure 2c), their initial state cannot be predicted precisely, perhaps ignoring the soil disturbance during the suction installation process of the caisson. The comparison of normalized results (in Figure 2d) shows that differential pressure, skirt wall friction and total vertical force can be forecasted approximatively at the end state (w = 0.015D). Moreover, this paper focuses on the uplift responses of caisson at the end state, so the calculation error can be ignored at the initial state.

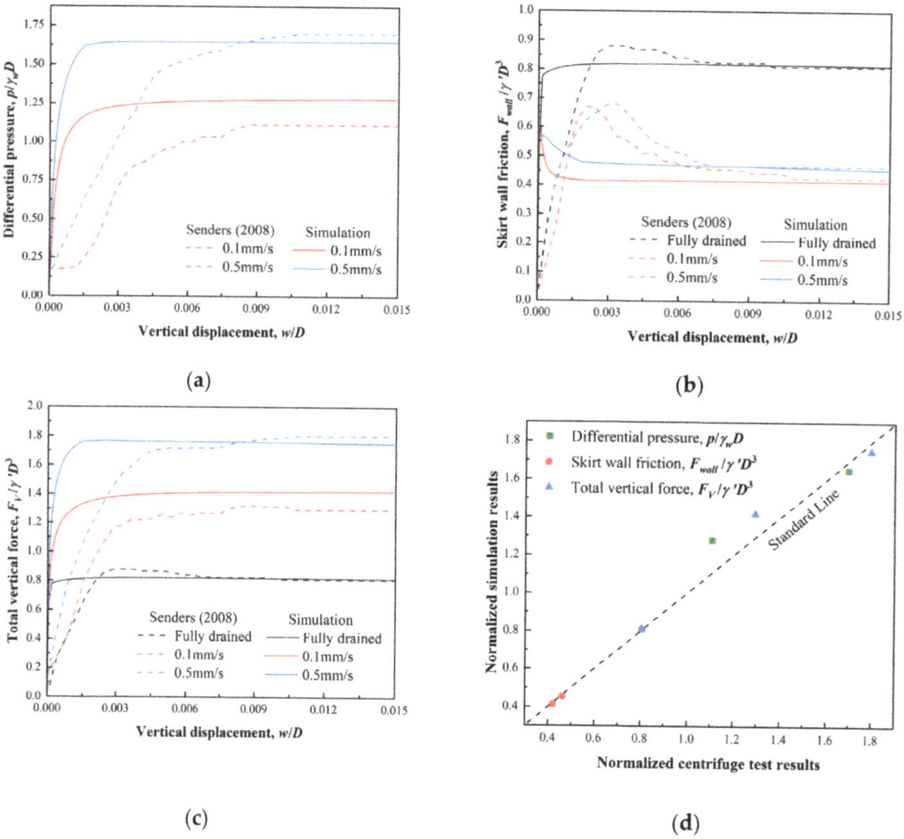

Figure 2. Comparisons between simulations and centrifuge tests: (**a**) differential pressure; (**b**) skirt wall friction; (**c**) total vertical force; and (**d**) comparison of normalized results at w = 0.015D.

2.3. Simulation Parameters

According to the test results of [22,35–37], the uplift responses of suction caisson are relevant with the pullout velocity when the vertical displacement exceeds 0.02D, and the caisson cannot meet the functional requirements at the same time. Therefore, the tensile loading when the vertical displacement of caisson reaches w_0 = 0.02D is regarded as its uplift capacity in this paper, and the responses of seepage, pore pressure, soil stress field, etc., are investigated later at this state.

The Dogger Bank sand in the North Sea is adopted in the next simulations with d_{10} = 0.087 mm, d_{60} = 0.174 mm, $e_{\max}$ = 0.865 and $e_{\min}$ = 0.597. The basic parameters of this paper's simulations are listed in Table 1. Blaker and Andersen [38,39] conducted a series of laboratory tests on Dogger Bank sand [40]. Based on the laboratory test results, Shen [15] calibrated the SANISAND model parameters of Dogger Bank sand, which are listed in Table 2.

Table 1. The basic parameters of this paper's simulations.

Properties	Values
Caisson	
Diameter, D (m)	10
Skirt length, L (m)	10
Skirt wall thickness, t_s (m)	0.1 ($t_s/D = 0.01$)
Friction coefficient of soil-structure, μ	0.5
Material property	Rigid body
Seabed soil	
Permeability coefficient, k_0 (m/s)	5×10^{-6}
Initial void ratio, e_s	0.658
Buoyant unit weight, γ' (kN/m^3)	10
Specific weight of water, γ_w (kN/m^3)	10
Earth pressure coefficient, K	0.45
Water element	
Element thickness, t_w (m)	0.2
Young's modulus, E_w (kPa)	10^{-5}
Poisson's ratio, n_w	0
Permeability coefficient, k_w (m/s)	1
Initial void ratio, e_w	0.658

Table 2. Parameters of the SANISAND model for Dogger Bank sand [15].

Parameter	Variable	Value
Elasticity	G_0	206
	v'	0.05
Critical state	M_c	1.4
	M_e	1.05
	λ_c	0.007
	e_{c0}	0.75
	e_{e0}	0.699
	ξ	0.7
Yield surface	m	0.05
Plastic modules	h_0	7
	c_h	0.98
	n^b	2.9
Dilatancy	A_0	1.6
	n^d	2
Fabric-dilatancy tensor	z_{max}	0
	c_z	0

3. Prediction of Differential Pressure Beneath Caisson Lid

3.1. Seepage Responses of Uplift Caisson

During the uplift progress of caisson, the pore pressure gradually accumulates, and a seepage field is formed in the soil domain. The enlargement of soil permeability results in the faster dissipation of pore pressure in the caisson and less accumulation of differential pressure beneath the caisson lid. Similarly, the decrease in pullout velocity generates longer pore pressure dissipation time and weakens its accumulation rate. The reduction of aspect ratio shortens the seepage length, causing a shorter dissipation time. Figure 3 shows that the differential pressure decreases with the soil permeability, and increases with the pullout velocity and aspect ratio, which are consistent with the previous descriptions.

The pore pressure fields for $v = 0.001$ mm/s, $v = 1$ mm/s and $v = 100$ mm/s are shown in Figure 4. For lower pullout velocity, the pore pressure inside the caisson gradually dissipates along the depth and its distribution is approximately stepped (see Figure 4a). However, the high pullout velocity generates larger cumulative pore pressure inside caisson, which is identical with Figure 3b, and the seepage field is developed to the deeper and farther seabed (see Figure 4b,c).

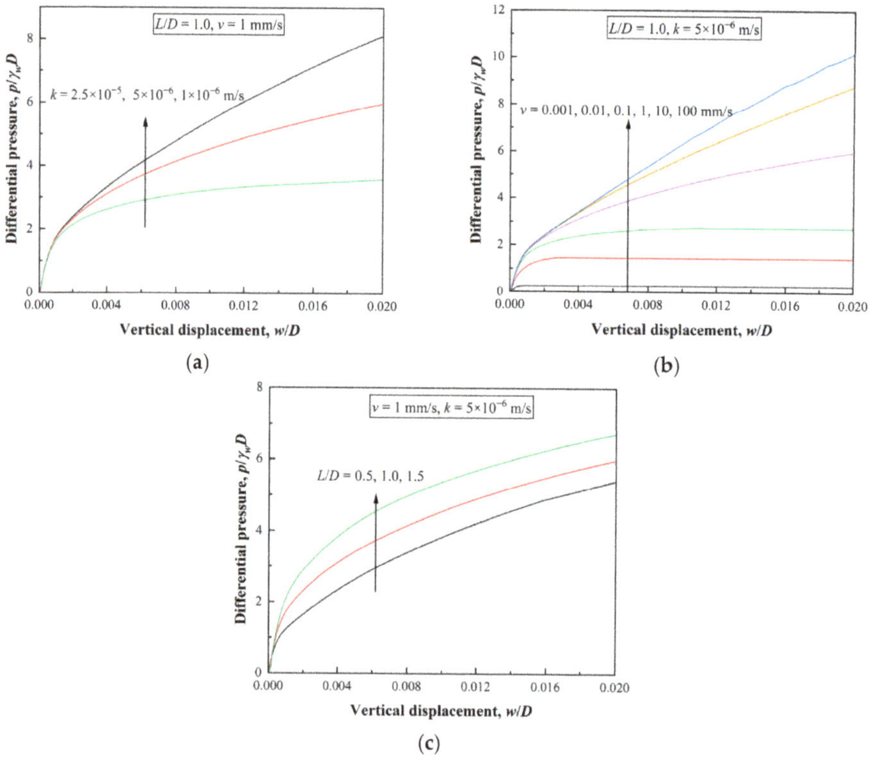

Figure 3. Differential pressure beneath caisson lid: (**a**) effects of permeability; (**b**) effects of pullout velocity; and (**c**) effects of aspect ratio (the caisson diameter is constant).

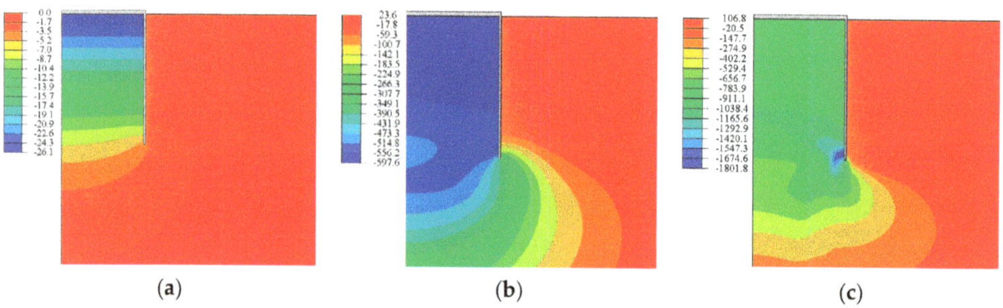

Figure 4. Pore pressure field (unit: kPa): (**a**) $k = 5 \times 10^{-6}$ & $v = 0.001$ mm/s; (**b**) $k = 5 \times 10^{-6}$ & $v = 1$ mm/s; and (**c**) $k = 5 \times 10^{-6}$ & $v = 100$ mm/s.

Figure 5 shows the soil relative displacement and flow direction for $v = 0.001$ mm/s, $v = 1$ mm/s and $v = 100$ mm/s. When the pullout velocity is 0.001 mm/s (in Figure 5a), the soil around skirt wall moves approximatively along the uplift direction. The soil flow is inapparent, presenting the characteristics of partial shear failure. As shown in Figure 5b, the high pullout velocity ($v = 1$ mm/s) results in the soil flow from the outside to the inside at the caisson tip and the direction of soil flow is accordant with the seepage, and the failure mechanism is closed to partial tensile failure. As the pullout velocity further increases ($v = 100$ mm/s, in Figure 5c), the scope of soil flow continuously expands, which presents the characteristics of overall reverse failure.

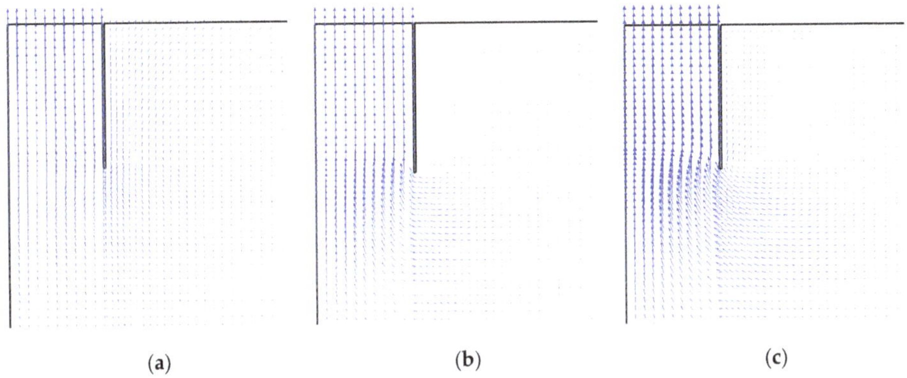

Figure 5. Soil displacement field: (**a**) $k = 5 \times 10^{-6}$ & $v = 0.001$ mm/s; (**b**) $k = 5 \times 10^{-6}$ & $v = 1$ mm/s; and (**c**) $k = 5 \times 10^{-6}$ & $v = 100$ mm/s.

3.2. *Differential Pressure Equation*

3.2.1. Equation Deduction

During a certain time step Δt, seepage will fill the void beneath the caisson lid with a certain amount of fluid [22], and the soil plug will be uplifted because of the differential pressure and internal skirt wall friction (see Figure 6). Therefore, the vertical pullout displacement ΔL is consisted of gap length ΔL_{gap} and plug height ΔL_{plug}. According to Senders [22], ΔL_{gap} can be expressed as:

$$\Delta L_{gap} = k i_{ave} \Delta t = k \frac{p}{\gamma_w s_{ave}} \Delta t \tag{1}$$

where i_{ave} is the average hydraulic gradient along the drainage path, p is the differential pressure beneath caisson lid. s_{ave} is the equivalent average seepage length, and the normalized seepage length s_{ave}/D is a function of L/D [12–16], which can be defined as:

$$\frac{s_{ave}}{D} = h(L/D) \tag{2}$$

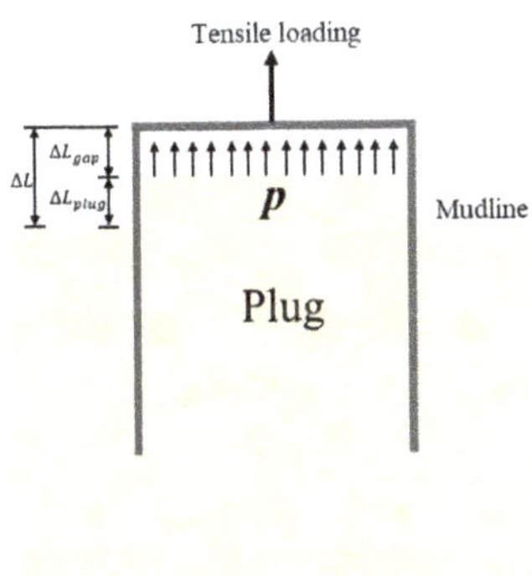

Figure 6. Displacement diagram of pullout suction caisson.

The normalized differential pressure $\frac{p}{\gamma_w D}$ can be expressed as:

$$\frac{p}{\gamma_w D} = \frac{\Delta L_{gap}}{\Delta t * k} * \frac{s_{ave}}{D} = \frac{\Delta L_{gap}}{\Delta L} * \frac{\Delta L}{\Delta t * k} * \frac{s_{ave}}{D} \tag{3}$$

Combining Equation (2) and Equation (3) can be expressed by:

$$\frac{p}{\gamma_w D} = \frac{\Delta L_{gap}}{\Delta L} * \frac{v}{k} * h\left(\frac{L}{D}\right) \quad (4)$$

Further, the gap length ΔL_{gap} is equal to vertical pullout displacement ΔL minus plug height ΔL_{plug}:

$$\frac{\Delta L_{gap}}{\Delta L} = 1 - \frac{\Delta L_{plug}}{\Delta L} \quad (5)$$

Considering the soil plug is constrained by internal skirt wall friction, plug height $\Delta L_{plug}/\Delta L$ is associated with the earth pressure coefficient K and interface friction angle of soil-structure δ. Furthermore, gap length $\Delta L_{gap}/\Delta L$ is also correlated with K and δ.

Based on the above, the determination of normalized differential pressure $p/\gamma_w D$ should consider the comprehensive impacts of pullout velocity v, permeability coefficient k, aspect ratio L/D, earth pressure coefficient K and interface friction angle of soil-structure δ. The shear stress acting on skirt wall is calculated by $\sigma_v' K tan\delta$, so the influence of K and δ is inseparable [20,21] and their impacts are reflected by the combination of $K tan\delta$.

3.2.2. Effect of Normalized Pullout Velocity

The differential pressure for different permeability coefficients ($0.2k_0$, k_0, $5k_0$) under various pullout velocity is shown in Figure 7a. The differential pressure is positively associated with pullout velocity and negatively relevant with permeability coefficient, which is consistent with the results of Section 3.1.

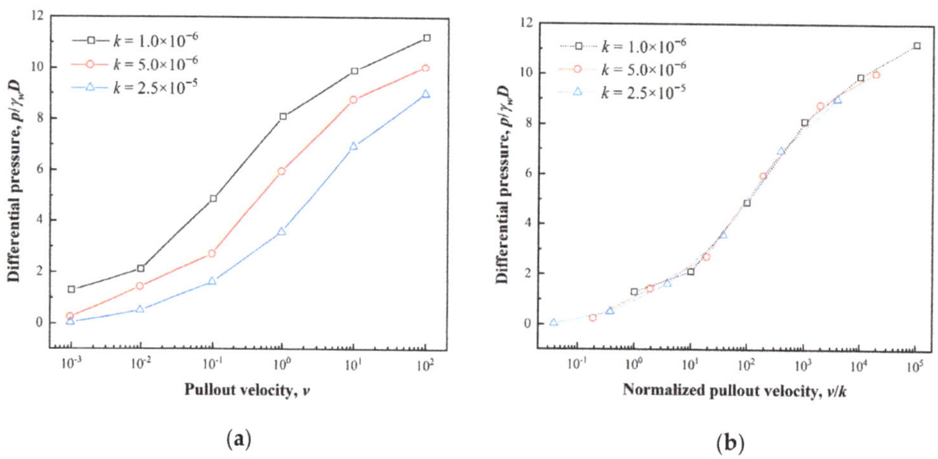

Figure 7. Differential pressure beneath the caisson lid (L/D = 1.0): (**a**) relationship with pullout velocity; and (**b**) relationship with normalized pullout velocity.

Based on Equation (1), the average hydraulic gradient along the drainage path can be defined as: $i_{ave} = p/(\gamma_w s_{ave})$. The caisson pullout velocity may be given roughly as: $v = k \cdot i_{ave}$ according to Darcy's law [17]. Therefore, the differential pressure beneath the lid can be expressed by

$$p = v/k \cdot \gamma_w s_{ave} \quad (6)$$

Normalized the pullout velocity by permeability coefficient (see Figure 7b), the differential pressure shows a high correlation with v/k and increases with it. This is coincident with Equation (6), which uniformly used Darcy's law to describe the seepage inside caisson. It is worth mentioning that Darcy's law applies to fluid flow in the range of Reynolds number $R_e < 10$ [41,42]. For the Dogger Bank sand used in this paper, its effective particle size d_{10}

is 0.087 mm, and the dynamic viscosity coefficient of water h is 1.01×10^{-6} m^2/s at a pressure of 101.325 kPa and a temperature of 20 °C [43]. The calculation of Reynolds number is defined as $R_e = \frac{v_s d_{10}}{\eta}$ [44], where v_s represents the fluid flow velocity. Therefore, the application range of Darcy's law can be expressed as: $\frac{v_s d_{10}}{\eta} < 10$, namely $v_s < 10\eta/d_{10} = 116$ mm/s. On account of the formation of gap between caisson lid and soil during the caisson being uplifted, the fluid flow velocity is always less than the pullout velocity. Under the limit state, the soil plug is completely pulled out, that is, soil plug ΔL_{plug} is equal to vertical pullout displacement ΔL and v_s is approximately equal to v. Therefore, the pullout velocity $v < 116$ mm/s can be regarded as a conservative condition to define the application scope of Darcy's law, and v/k is less than 2.32×10^4 for the Dogger Bank sand used in this paper.

Considering the large fluid flow velocity where Darcy's law is not applied, a wider range of uplift velocity is selected and the normalized pullout velocity v/k is from 10^{-5} to 10^8. The changes in differential pressure are shown in Figure 8, and the calculation data is fitted by hyperbolic tangent curve, which can be expressed as:

$$f(v/k) = \begin{cases} e^{8.61*\tanh(0.3\,\lg\frac{v}{k}+0.87)-6.07}, & v/k \le 10^4 \\ e^{2.5*\tanh(0.43\,\lg\frac{v}{k}-0.09)-0.025}, & v/k > 10^4 \end{cases} \quad (7)$$

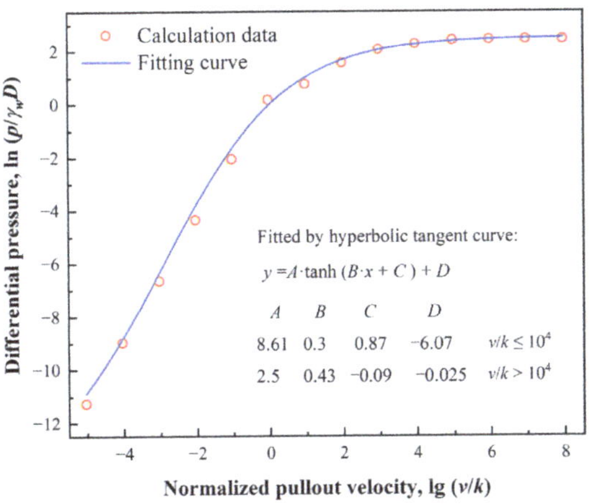

Figure 8. Differential pressure for different normalized pullout velocity ($L/D = 1.0$).

3.2.3. Effect of Aspect Ratio

From Section 3.1, the enlargement of aspect ratio extends the seepage length, so the differential pressure increases with it. For v/k from 10^{-5} to 10^8 and $L/D = 0.5, 1.0, 1.5$ & $m = 0.5$, the variation in differential pressure is shown in Figure 9a. For different normalized pullout velocity, the correlation between differential pressure and aspect ratio is shown in Figure 9b.

The differential pressure is seldomly generated at a low pullout velocity corresponding to $v/k \le 10^{-1}$. As the pullout velocity increases, the differential pressure gradually accumulates corresponding to $10^{-1} < v/k \le 10^4$. According to the analysis of Section 3.2.2, Darcy's law is still satisfied at this time. With the further increase of pullout velocity, the enlargement of differential pressure is limited corresponding to $v/k > 10^4$. Meanwhile, Darcy's law is not suitable for describing the seepage response inside caisson. The bottom and top boundaries (see Figure 9b) are defined as the separatrices of different drainage conditions [24]. Therefore, the soil inside caisson is closed to fully drained condition for $v/k \le 10^{-1}$ [45], partially drained for $10^{-1} < v/k \le 10^4$ and undrained for $v/k > 10^4$.

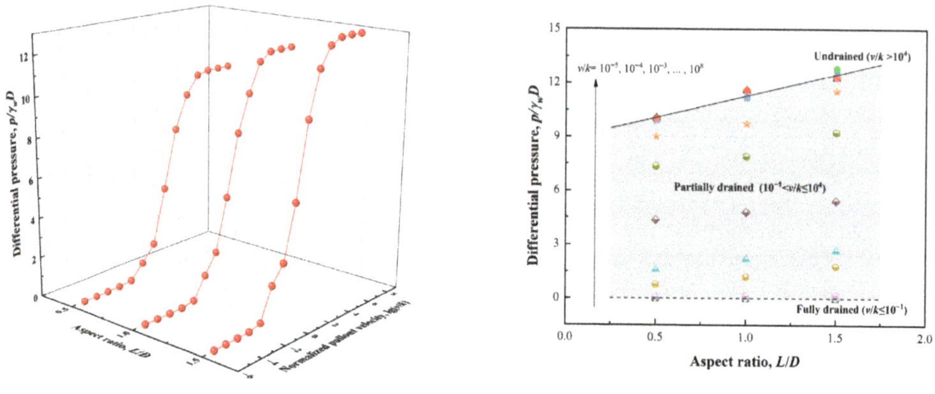

(a) (b)

Figure 9. Differential pressure for different aspect ratio and normalized pullout velocity: (**a**) in three dimensions; and (**b**) in two dimensions.

Taking the results of $L/D = 1.0$ as a benchmark, the calculation data is fitted by piecewise linear equation to consider the effects of aspect ratio, which can be expressed by:

$$g(L/D) = \begin{cases} 1, & v/k \leq 10^{-1} \\ 0.236L/D + 0.764, & 10^{-1} < v/k \leq 10^{4} \\ 0.212L/D + 0.788, & 10^{4} < v/k \end{cases} \quad (8)$$

3.2.4. Effect of Ktanδ

The combined influence of *Ktanδ* is reflected by changing the friction coefficient of soil-structure, which is associated with interface friction angle. According to [46], $\delta = 10°$ (corresponding to $\mu = 0.176$) and $d = 35°$ (corresponding to $\mu = 0.700$) can be treated as the relative smooth and rough soil-structure interface, respectively. Therefore, for v/k from 10^{-5} to 10^{8} and $m = 0.1, 0.3, 0.5, 0.7$ and $L/D = 1.0$, the variation of differential pressure is shown in Figure 10a. For different normalized pullout velocity, the correlation between differential pressure and friction coefficient of soil-structure is shown in Figure 10b.

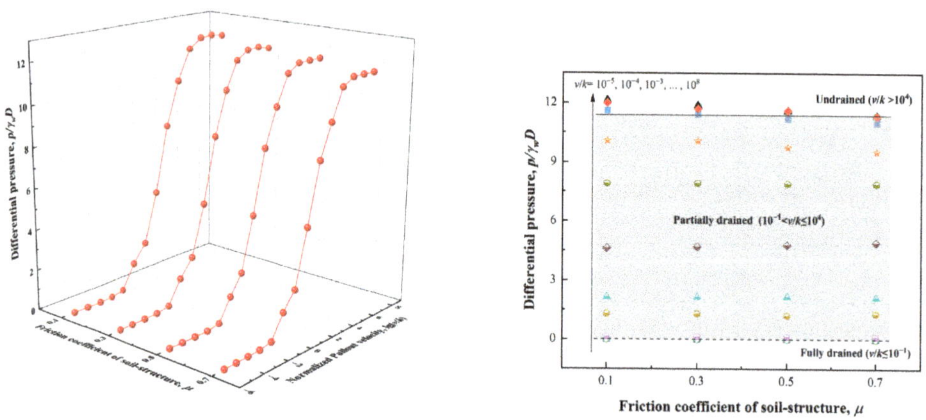

(a) (b)

Figure 10. Differential pressure for different friction coefficient of soil-structure and normalized pullout velocity: (**a**) in three dimensions; and (**b**) in two dimensions.

As normalized pullout velocity v/k increases, the differential pressure enlarges and the soil inside the caisson presents various drainage conditions, but the impacts of friction coefficient on seepage is not significant. Therefore, the effects of $Ktan\delta$ can be expressed as:

$$j(K, \delta) = 1 \tag{9}$$

3.2.5. Verification of Differential Pressure Equation

Combined with the above, the differential pressure equation can be defined as:

$$p = f(v/k) \cdot g(L/D) \cdot j(K, \delta) * \gamma_w D \tag{10}$$

The curve of differential pressure equation (DPE curve) is also divided into three segments corresponding to various drainage conditions.

Figure 11 shows the comparisons of DPE curve with three groups of test results. In Section 2.3, the initial void ratio of silica sand used in [22] centrifuge tests is 0.7357, soil permeability coefficient is 2.3×10^{-4} m/s and aspect ratio L/D is 1.0. The DPE curve coincides well with the two tests (v = 0.1, 0.5 mm/s).

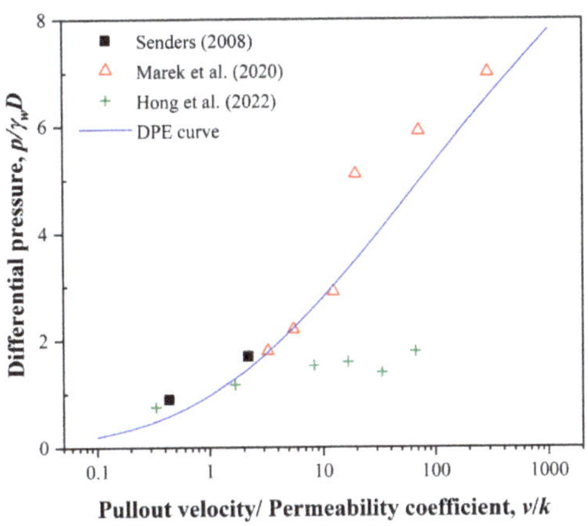

Figure 11. Comparisons between test results and DPE curve.

Marek [47] conducted a series of experiment investigations for caisson foundations in sand subjected to rapid uplift. The initial void ratio of Lubiatowo sand used in Marek (2020) experiments is 0.64, soil permeability coefficient is 1.54×10^{-3} m/s and aspect ratio L/D is 1.0. The pullout velocity combines 5, 8.5, 19, 30, 110 and 450 mm/s. The DPE curve presents a better prediction for differential pressure.

Hong [19] performed a series of 1 g tests to research the effects of drainage conditions on the pullout characteristics of caisson foundations in sand. The initial void ratio of sand used in Hong [19] tests is 0.826, soil permeability coefficient is 3×10^{-4} m/s and aspect ratio L/D is 1.0. The pullout velocity includes 0.1, 0.5, 2.5, 5, 10 and 20 mm/s. The DPE curve coincides well with Hong [19] tests with v = 0.1 mm/s and 0.5 mm/s. As the pullout velocity increases, the change in differential pressure is slight and mismatches the prediction results of DPE curve. The possible reason is that the scale effect of seepage cannot be satisfied in a 1g test, causing the impacts of differential pressure to not be amplified into the prototype.

4. Calculation Method of Uplift Capacity

4.1. The Uplift Capacity Mechanism of Suction Caisson

For a certain seabed, there are little differential pressure beneath caisson lid and tiny soil plug inside caisson when the pullout velocity is small [48–50]. As the pullout velocity increases, the differential pressure gradually accumulates and soil plug expands. Neglecting the interaction between caisson and plug, the overall bearing mode is shown in Figure 12a, including uplift loading F_{tot}, external skirt wall friction F_{fri_out}, gravity W_{plug} + W_{cai} and bottom force F_{bottom}, which can be expressed by:

$$F_{tot} = F_{fri\_out} + W_{plug} + W_{cai} + F_{bottom} \tag{11}$$

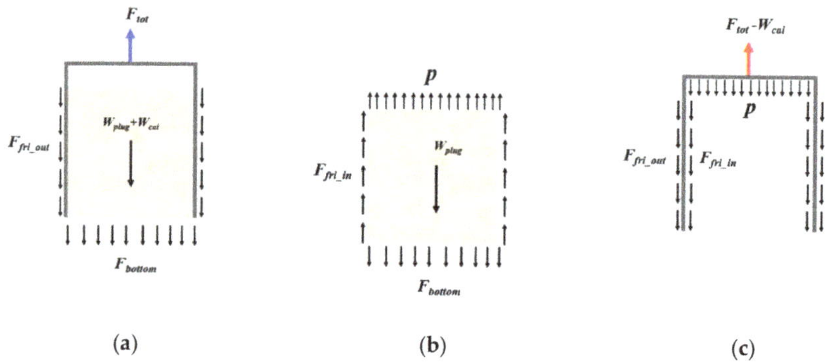

Figure 12. Bearing capacity mechanism: (**a**) overall bearing mode; (**b**) soil plug module; and (**c**)suction caisson module.

Dividing the overall bearing mode into soil plug and suction caisson modules, their force modes are shown in Figure 12b,c, which can be expressed as:

$$p * \pi D_i^2/4 + F_{fri\_in} = W_{plug} + F_{bottom} \tag{12}$$

$$F_V = F_{tot} - W_{cai} = p * \pi D_i^2/4 + F_{fri\_in} + F_{fri\_out} \tag{13}$$

where p is the differential pressure beneath the lid and can be calculated by Equation (10). F_{fri_in} is the internal skirt wall friction.

The calculations of internal and external skirt wall friction can be expressed as:

$$F_{fri\_in} = \int_0^L \sigma_v' dz \cdot (Ktan\delta)_i(\pi D_i) \tag{14}$$

$$F_{fri\_out} = \int_0^L \sigma_v' dz \cdot (Ktan\delta)_o(\pi D_o) \tag{15}$$

where σ_v' is the vertical effective stress. D_i and D_o are the internal and external diameters of caisson, respectively.

The differential pressure contributes to the uplift capacity of suction caisson, and the skirt wall friction is also changed due to the seepage field in soil domain. The uplift capacity F_V is normalized by the fully drained uplift resistance, and the relationship between the normalized uplift capacity F_V' against the normalized pullout velocity v/k is shown in Figure 13. Under fully drained condition, the uplift capacity is mainly contributed by skirt wall friction. As the pullout velocity increases, the seepage effect is gradually prominent, and the differential pressure accumulates beneath the caisson lid. The uplift capacity can reach 10 times or more of the fully drained resistance, showing an undrained response. Further, with the decrease in interface friction coefficient of soil-structure, the uplift capacity gradually enlarges. Therefore, the design of suction caisson is conservative when the drainage condition is not considered [25].

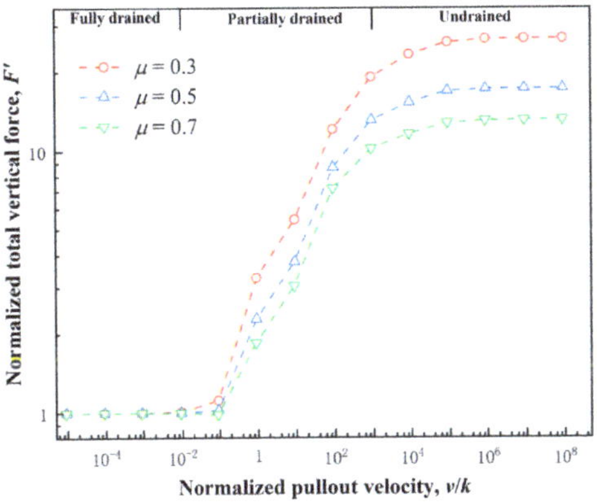

Figure 13. Normalized uplift capacity against the normalized pullout velocity.

4.2. Fully Drained Capacity

4.2.1. Soil Stress State

For the fully drained condition, the effects of differential pressure can be ignored, and the pivotal point is how to determine the soil effective stress acting on the skirt wall, namely σ_v' [20,21]. Extracting the soil effective stress around skirt wall along the depth (in Figure 14), the soil stress at the pullout state is different from in situ soil stress. One possible factor is that the soil around skirt is disturbed due to the relative movement of the soil-structure interface, leading to the variations of soil stress field.

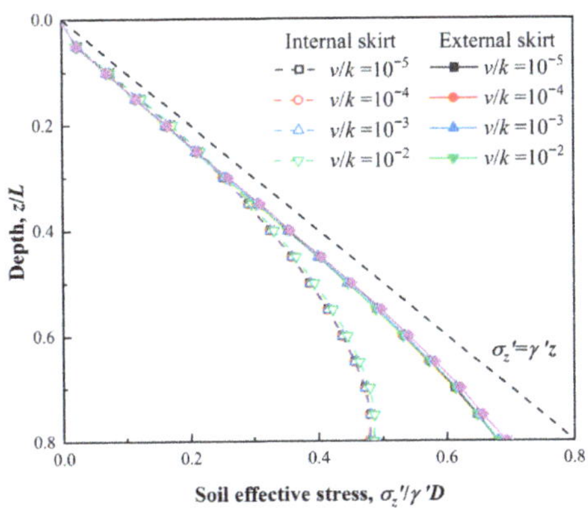

Figure 14. Soil effective stress along the depth.

4.2.2. Soil Stress Release Model

Houlsby [20,21] proposed an analytical model to reflect the stress reduction effects of seabed soil during caisson being pulled out. For caisson internal side, assuming that the internal skirt friction results in uniform reduction of soil vertical stress, it generates proportional decrease in soil-skirt normal contact stress [14]. Based on this assumption, the vertical soil stress on internal skirt is calculated by:

$$\frac{d\sigma_v'}{dz} = \gamma' - \frac{\sigma_v'(Ktan\delta)_i(\pi D)_i}{\pi D_i^2/4} = \gamma' - \frac{4\sigma_v'(Ktan\delta)_i}{D_i} \quad (16)$$

$$\int_0^L \sigma_v' dz = \gamma' Z_i^2 [\exp\left(-\frac{L}{Z_i}\right) - 1 + \frac{L}{Z_i}] \quad (17)$$

$$Z_i = \frac{D_i}{4(Ktan\delta)_i} \tag{18}$$

For caisson external side, it is supposed that there is a uniform soil stress release region within a certain range outside the skirt. Assuming that the stress release region is in a cylindrical soil ring outside the skirt, the outer diameter of release region is defined as $D_m = mD_o$ (m is the outer diameter coefficient), and the vertical soil stress on external skirt is calculated by:

$$\frac{d\sigma'_v}{dz} = \gamma' - \frac{\sigma'_v(Ktan\delta)_o(\pi D)_o}{\pi(D_m^2 - D_o^2)/4} = \gamma' - \frac{4\sigma'_v(Ktan\delta)_o}{(m^2-1)D_o} \tag{19}$$

$$\int_0^L \sigma'_v dz = \gamma' Z_o^2 [\exp\left(-\frac{L}{Z_o}\right) - 1 + \frac{L}{Z_o}] \tag{20}$$

$$Z_o = \frac{D_o(m^2-1)}{4(Ktan\delta)_o} \tag{21}$$

Considering the effects of normalized pullout velocity v/k and $Ktan\delta$ on the range of stress release region, the soil effective stress (caisson aspect ratio L/D = 1.0; soil depth z = 5.0 m) along the radial direction of caisson is shown in Figure 15. Under fully drained, the change in normalized pullout velocity v/k has negligible impacts on the soil stress state (in Figure 15a). As shown in Figure 15b, the reduction of soil stress increases with the friction coefficient of soil-structure, presenting that the degree of soil disturbance enlarges, but the range of soil release region does not change. Next, the soil stress presents an approximately linear attenuation within the stress release region. At x/D = 0.7 in Figure 15, it is found that the curve of soil effective stress turns and soil stress no longer decreases along radial direction, and the position of x/D = 0.7 is defined as the outer boundary of stress release region. Therefore, the outer diameter of stress release region is defined as $D_m = 1.4D_o$ (m = 1.4) for L/D = 1.0 & z = 5.0 m.

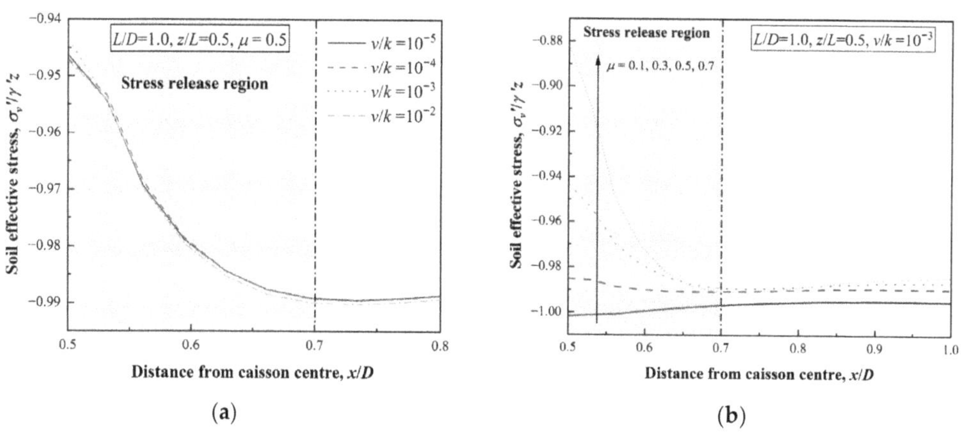

Figure 15. Soil effective stress along the caisson diameter: (**a**) considering the effects of v/k; and (**b**) considering the effects of $Ktan\delta$.

Figure 16 shows the soil effective stress for L/D = 0.5 and z = 1.0, 3.0 m, L/D = 1.0 and z = 1.0, 3.0, 5.0, 7.0 m and L/D = 1.5 and z = 1.0, 3.0, 5.0, 7.0, 10.0, 12.0, 14.0 m. Referring to the above method for defining the outer diameter, the gray area in Figure 16 is the stress release region. The outer diameter coefficient m is approximately linearly related to depth z, which is different from the assumption of Houlsby [21], and m is expressed as:

$$m = 0.08z + 1.0 \tag{22}$$

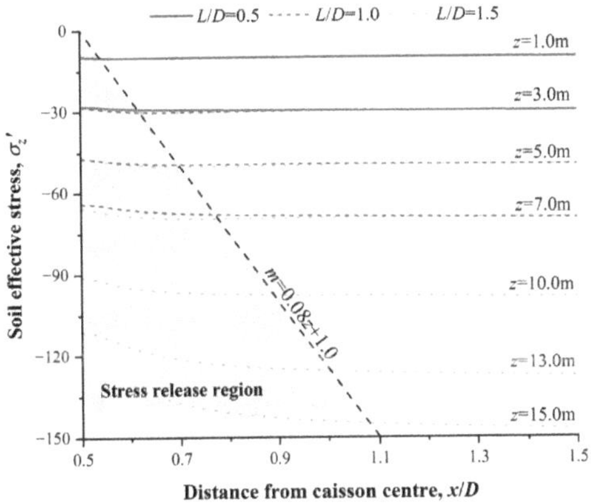

Figure 16. Soil effective stress along the caisson diameter at different depth.

Combined with the previous analysis and the assumption of Houlsby [21], it is further supposed: (1) there is a stress disturbance zone that expands along the depth, and the outer diameter coefficient m is linearly related to depth z; and (2) the vertical earth pressure within the stress release region is calculated according to the soil stress at $x = (D_o + D_m)/2$.

The diagram of stress release model proposed by this paper is shown in Figure 17. For the soil slice in the range $z = h_i$ to $z = h_i + \Delta h$ (Δh is small enough, and m is regarded as a constant value in the range of Δh), its vertical force equilibrium inside the caisson remains the same (see Equations (16)~(18)), and that outside the caisson can be expressed as:

$$\sum \int_{h_i}^{h_i+\Delta h} \sigma_v' dz = \gamma' Z_o^2 [\exp\left(-\frac{h_i+\Delta h}{Z_o}\right) - \exp\left(-\frac{h_i}{Z_o}\right) + \frac{\Delta h}{Z_o}] \tag{23}$$

where Z_o and m can be referred to Equations (21) and (22), respectively.

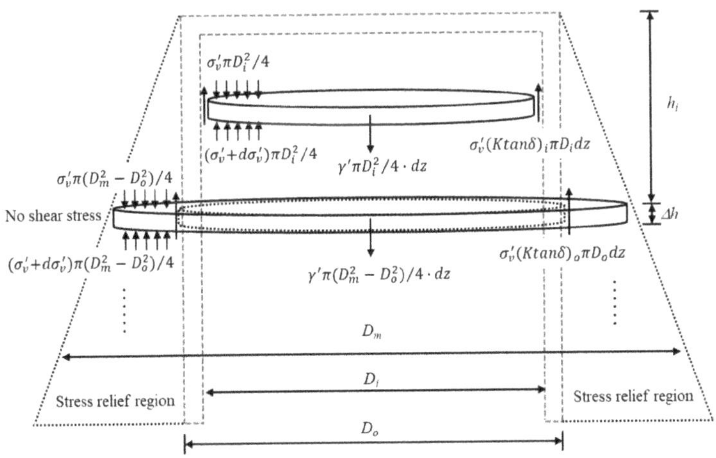

Figure 17. Soil effective stress along the caisson diameter at different depth.

4.2.3. Verification of Soil Stress Release Model

Combined with Equations (14)~(23), the analytical solution of total skirt wall friction is shown in Figure 18, which is compared with the simulation solution and analytical solution proposed by Houslby [21]. The analytical solution of this paper, which considers the variation of the stress release range along the depth, is closer to the simulation solution than adopting fixed outer diameter coefficient (m = 1.2, 1.5 in Figure 19) proposed by Houslby et al. (2005) [21]. Moreover, the analytical solution of Houslby et al. (2005) [21] increases with the value of m, so its accuracy depends on the determination of stress release range.

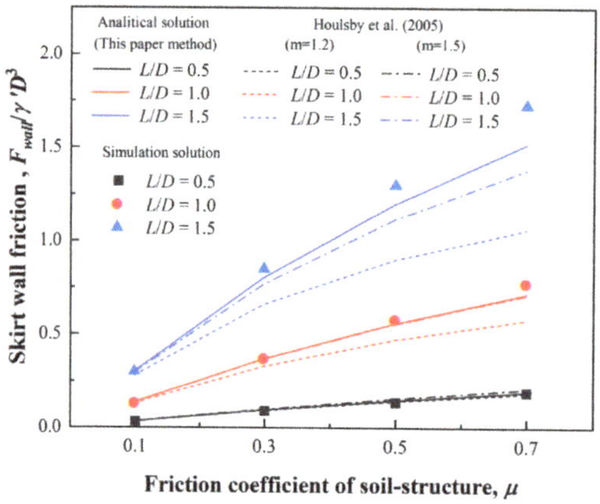

Figure 18. Comparison of simulation solution with analytical solution (fully drained).

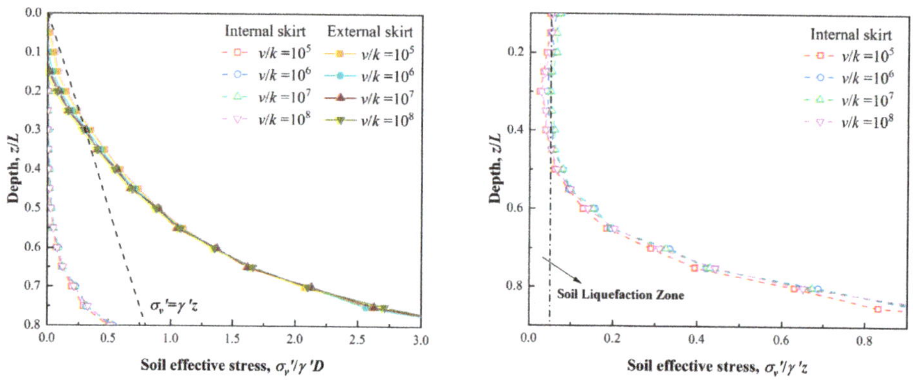

(**a**) (**b**)

Figure 19. Soil effective stress around caisson along depth (**a**) internal skirt and external skirt; (**b**) internal skirt.

In Figure 18, the difference value between analytical solution of this paper and simulation solution obviously enlarges with the friction coefficient of soil-structure, and total skirt wall friction is underestimated by 12.06% at L/D = 1.5 & m = 0.7. Under the constraint of surrounding soil, the soil dilatancy caused by interface shear stress generates an increase in the normal effective stress. However, this dilatancy effect is affected by the interface friction angle between the soil and skirt wall. As the sand reaches the interface limit friction force, slippage first occurs at the interface between the soil grains and the skirt wall, and relative sliding and dilatancy no longer occurs inside the sand [14]. As the interface roughness of soil-structure increases, the soil dilatancy intensifies before the sand reaches the interface limit friction force, causing the increment of soil stress produced by soil dilatancy to also enlarge. The stress release model neglects the increase in soil stress caused by soil dilatancy, so the difference enlarges with the friction coefficient.

According to Section 2.2, Senders [22] conducted a group of centrifuge test under fully drained. The skirt wall friction measured in centrifuge test is compared with the analytical solution, which is summarized in Table 3. For the analytical solution of Houslby et al. (2005) [21], although the calculation results can be closer to the centrifuge test by adjusting the value of m, it is difficult to obtain the accurate value of m for every project, which brings difficulty to the engineering application.

Table 3. Comparison of centrifuge test (Sender [22]) with analytical solution (fully drained).

	Calculation Method	Skirt Wall Friction (kN)	Relative Error
Centrifuge test	Senders [22]	1750	-
Analytical solution	Houlsby et al. (2005) $m = 1.2$	1249.37	−28.6%
	Houlsby et al. (2005) $m = 1.5$	1543.75	−11.8%
	Stress release model this paper	1478.27	−15.5%
	(Relative error less than 0 indicates underestimation)		

In this paper, the outer diameter coefficient m is defined as a depth dependent parameter and its expression is given in Equation (22). The analytical solution of this paper method is 1478.27 kN with an error of −15.5% (in Table 3). Although the result is conservative, it facilitates the application of the project and improves the prediction accuracy of uplift capacity.

4.3. Undrained Capacity

In sand, two cases need to be considered for tensile loading: slow (fully drained) loading and rapid (undrained) loading [27]. For fully drained condition, the calculation method of uplift capacity is introduced in Section 4.2. Then, for the undrained condition, the uplift resistance contributed by differential pressure enlarges the uplift capacity of suction caisson, but the soil effective stress inside and outside the caisson is also being changed. Figure 16 shows the soil effective stress around skirt wall along depth. The soil effective stress outside caisson is enhanced due to seepage, but the large pore pressure in the caisson leads to soil liquefaction. The reduction of soil effective stress is severe, resulting in nearly half of the soil (0~0.5L, Figure 19b) being liquefied (soil strength loss reaches to 95% [51,52]).

Assuming that the increase in outer skirt wall friction due to seepage force is the same as the decrease in inner that [23,24], a simplified calculation method of uplift capacity [27] is expressed as:

$$F_V = p * \pi D_i^2/4 + \frac{\gamma' h^2}{2}[(Ktan\delta)_i(\pi D_i) + (Ktan\delta)_o(\pi D_o)] \tag{24}$$

which can be simplified as:

$$F_V = p * \pi D^2/4 + 2\pi D\frac{\gamma' h^2}{2}Ktan\delta \tag{25}$$

The comparison of total vertical force is shown in Table 4 for L/D = 1.0. The relative error is less than 5%, presenting good prediction accuracy of uplift capacity. Under undrained condition, the differential pressure has a pivotal effect on anti-pullout of caisson (see Figure 13), and its evaluation is the key to the calculation of suction bucket uplift capacity. The influence of friction coefficient is reflected in skirt wall friction, whose contribution is smaller than differential pressure. Therefore, the influence of friction coefficient on uplift capacity is insignificant in Table 4.

Table 4. Comparison of simulation with analytical solution.

Friction Coefficient	Total Vertical Force (kN)		Relative Error
	Simulation Solution	Analytical Solution	
0.1	96,489	91,813	−4.85%
0.3	98,063	93,227	−4.93%
0.5	99,541	94,640	−4.92%
0.7	100,368	96,054	−4.30%

5. Conclusions

Considering the soil drainage conditions during caisson being pulled out, the fully drained, partially drained and undrained are divided. An equation is proposed to evaluate the development of differential pressure beneath the caisson lid, which can quantify its contribution to uplift capacity under undrained. Further, the calculation method considering soil stress release is proposed to calculate the uplift capacity under fully drained. Key conclusions from this study are summarized below:

(1) During caisson being pulled out, the differential pressure is positively associated with normalized pullout velocity v/k and aspect ratio L/D, but the impacts of lateral earth coefficient K and interface friction angle of soil-structure δ on seepage is not significant. When the normalized pullout velocity v/k is less than 10^{-1}, the differential pressure is seldomly generated, presenting the fully drained response. The enlargement of differential pressure is limited and the soil inside caisson is closed to undrained condition for $v/k > 10^4$.

(2) Under fully drained condition, the uplift capacity of suction caisson is basically undertaken by skirt wall friction. The soil effective stress around skirt wall has been reduced due to the relative slippage between soil and caisson. On account of the stress reduction, a stress release model, considering the variation of stress release range along the depth, is proposed to calculate the skirt wall friction. Compared with the previous method of fixing the range of stress release region, it facilitates the application of the project and improves the prediction accuracy of uplift capacity.

(3) Under undrained condition, the uplift capacity of suction caisson can reach 10 times or more of the fully drained resistance due to the differential pressure acting on the caisson lid. However, the reduction of soil effective stress inside caisson is severe, and nearly half of the soil is liquefied. The differential pressure has a pivotal effect on anti-pullout of caisson, and its impact is calculated based on the proposed differential pressure equation. Then, a simplified method is proposed to calculate the uplift capacity, which is composed of skirt wall friction and differential pressure, showing good prediction accuracy.

Author Contributions: Conceptualization, C.X.; Methodology, M.X.; Software, M.X.; Validation, C.X.; Formal analysis, C.X. and H.J.; Investigation, D.S.; Data curation, C.X.; Writing—original draft, M.X.; Writing—review & editing, S.R. All authors have read and agreed to the published version of the manuscript.

Funding: This work was financed by the National Natural Science Foundation of China (51779220) and the Natural Science Foundation of Zhejiang Province (LR22E080005). The authors are grateful for the receipt of this funding.

Institutional Review Board Statement: Not applicable.

Informed Consent Statement: Not applicable.

Data Availability Statement: All relevant data is presented in the manuscript.

Conflicts of Interest: The authors declare no conflict of interest.

References

1. Pérez-Collazo, C.; Greaves, D.; Iglesias, G. A review of combined wave and offshore wind energy. *Renew. Sustain. Energy Rev.* **2015**, *42*, 141–153. [CrossRef]
2. Guo, Z.; Yu, L.Q.; Wang, L.Z.; Bhattacharya, S.; Nikitas, G.; Xing, Y. Model tests on the long-term dynamic performance of offshore wind turbines founded on monopiles in sand. *J. Offshore Mech. Arct. Eng.* **2015**, *137*, 041902. [CrossRef]
3. Abhinav, K.A.; Saha, N. Stochastic response of jacket supported offshore wind turbines for varying soil parameters. *Renew. Energy* **2017**, *101*, 550–564. [CrossRef]
4. Zhou, W.J.; Wang, L.Z.; Guo, Z.; Zhang, Y.R.; Rui, S.J. Numerical model for suction caisson under axial cyclic loadings. *Ocean Eng.* **2021**, *240*, 109956. [CrossRef]
5. Wang, X.F.; Zeng, X.W.; Li, J.L. Vertical performance of suction bucket foundation for offshore wind turbines in sand. *Ocean Eng.* **2019**, *180*, 40–48. [CrossRef]
6. Achmus, M.; Abdel-Rahman, K.; Schaefer, D.; Kuo, Y.S.; Chung, C.Y.; Tseng, Y.H. Capacity degradation method for driven steel piles under axial cyclic loadings. In Proceedings of the International Society of Offshore and Polar Engineers, San Francisco, CA, USA, 25–30 June 2017.
7. Bhattacharya, S.; Nikitas, G.; Arany, L.; Nikitas, N. Soil-structure interactions (SSI) for offshore wind turbines. *Eng. Technol. Ref.* **2017**, *24*, 16.
8. Guo, Z.; Wang, L.Z.; Yuan, F. Set-up and pullout mechanism of suction caisson in a soft clay seabed. *Mar. Georesour. Geotechnol.* **2014**, *32*, 135–154. [CrossRef]
9. Iskander, M.; El-Gharbawy, S.; Olson, R. Performance of suction caissons in sand and clay. *Can. Geotech. J.* **2002**, *39*, 576–584. [CrossRef]
10. Patel, S.K.; Singh, B. A parametric study on the vertical pullout capacity of suction caisson foundation in cohesive soil. *Innov. Infrastruct. Solut.* **2019**, *4*, 1. [CrossRef]
11. Guo, Z.; Jeng, D.; Guo, W.; He, R. Simplified approximation for seepage effect on penetration resistance of suction caissons in sand. *Ships Offshore Struct.* **2017**, *12*, 980–990. [CrossRef]
12. Mana, D.S.K.; Gourvenec, S.; Randolph, M.F. Numerical Modelling of Seepage beneath Skirted Foundations Subjected to Vertical Uplift. *Comput. Geotech.* **2014**, *55*, 150–157.
13. Thieken, K.; Achmus, M.; Schroder, C. On the Behavior of Suction Buckets in Sand under Tensile Loads. *Comput. Geotech.* **2014**, *60*, 88–100. [CrossRef]
14. Shen, K.; Zhang, Y.; Klinkvort, R.T.; Sturm, H.; Jostad, H.P.; Sivasithamparam, N.; Guo, Z. Numerical simulation of suction bucket under vertical tension loading. In Proceedings of the 8th International Conference on Offshore Site Investigation and Geotechnics, London, UK, 12–14 November 2017; Volume 1, pp. 488–497.
15. Shen, K.; Zhang, Y.; Wang, K.; Wang, B.; Zhao, X. Effect of partial drainage on the pullout behaviour of a suction bucket foundation. *Europ. J. Environ. Civil Eng.* **2021**, *26*, 1–29.
16. Wang, X.X.; Yang, Q.; Tang, X.W. Responses of suction buckets subjected to sustained vertical uplift. *Mar. Georesources. Geotechnology* **2020**, *40*, 36–51. [CrossRef]
17. Moghaddam, F.; Imam, R. The seepage study of transient caisson pull-out in sand. *Ocean Eng.* **2022**, *258*, 111804.
18. Vaitkunaite, E. Physical Modelling of Bucket Foundations Subjected to Axial Loading. Ph.D. Thesis, Aalborg University, Aalborg, Denmark, 2016.
19. Hong, S.; Vicent, S.; Gu, K.Y.; Kim, S.R. Effect of drainage condition on the pullout characteristics of bucket foundations in sand. *Ocean Eng.* **2022**, *260*, 111994.
20. Houlsby, G.T.; Byrne, B.W. Design procedures for installation of suction caissons in sand. *Geotech. Eng.* **2005**, *158*, 135–144. [CrossRef]
21. Houlsby, G.T.; Kelly, R.B.; Byrne, B.W. The tensile capacity of suction caissons in sand under rapid loading. In Proceedings of the International Symposium on Frontiers in Offshore Geomechanics, Perth, Australia, 19–21 September 2005; pp. 405–410. [CrossRef]
22. Senders, M. Suction Caissons in Sand as Tripod Foundations for Offshore Wind Turbines. Ph.D. Thesis, The University of Western Australia, Perth, Australia, 2008.
23. Vicent, S.; Kim, S.R.; Bong, T. Effect of loading rate on the pullout capacity of offshore bucket foundations in sand. *Ocean Eng.* **2020**, *210*, 107427.
24. Vicent, S.; Hong, S.; Bong, T.; Kim, S.R. Effects of embedment depth on the pull-out capacity of bucket foundations in sand. *Ocean Eng.* **2021**, *237*, 109643.
25. DNV. *Support Structures for Wind Turbines (DNVGL-ST-0126)*; Det Norske Veritas: Bærum, Norway, 2018.
26. *ISO 19901-4:2003*; Specific Requirements for Offshore Structures, Part 4-Geotechnical and Foundation Design Considerations. International Organisation for Standardisation: Geneva, Switzerland, 2014.
27. OWA. Suction Installed Caisson Foundations Design Guidelines for Offshore Wind. Offshore Wind Accelerator. 2019. Available online: https://www.engins.org/event/owa-suction-installed-caisson-foundations-design-guidelines-for-offshore-wind/ (accessed on 19 December 2022).
28. Guo, Z.; Jeng, D.S.; Guo, W.; Wang, L.Z. Failure mode and capacity of suction caisson under inclined short-term static and one-way cyclic loadings. *Mar. Georesour. Geotechnol.* **2018**, *36*, 52–63. [CrossRef]

29. *Systemes, Dassault, Abaqus 2017 Documentation*; Dassault Systèmes Simulia Corp: Providence, RI, USA, 2017.
30. Cao, J.; Phillips, R.; Popescu, R. Numerical analysis of the behavior of Suction Caissons in clay. *Int. J. Offshore Polar Eng.* **2003**, *13*, 154–159.
31. Dafalias, Y.F.; Manzari, M.T. Simple plasticity sand model accounting for fabric change effects. *J. Eng. Mech.* **2004**, *130*, 622–634. [CrossRef]
32. Dafalias, Y.F.; Papadimitriou, A.G.; Li, X.S. Sand plasticity model accounting for inherent fabric anisotropy. *J. Eng. Mech.* **2004**, *130*, 1319–1333. [CrossRef]
33. Dafalias, Y.F.; Taiebat, M. SANISAND-Z: Zero elastic range sand plasticity model. *Geotechnique* **2016**, *66*, 999–1013. [CrossRef]
34. Tran, M.N. Installation of Suction Caissons in Dense Sand and the Influence of Silt and Cemented Layers. Ph.D. Thesis, The University of Sydney, Sydney, Australia, 2005.
35. Byrne, B.W.; Houlsby, G.T. Experimental investigations of the response of suction caissons to transient vertical loading. *J. Geotech. Geoenviron. Eng.* **2002**, *128*, 926–939.
36. Byrne, B.W.; Houlsby, G.T. Experimental investigations of the response of suction caissons to transient combined loading. *J. Geotech. Geoenviron. Eng.* **2004**, *130*, 240–253.
37. Zhu, B.; Dai, J.L.; Kong, D.Q.; Feng, L.Y.; Chen, Y.M. Centrifuge modelling of uplift response of suction caisson groups in soft clay. *Can. Geotech. J.* **2020**, *57*, 1294–1303. [CrossRef]
38. Blaker, Ø.; Andersen, K.H. Shear strength of dense to very dense Dogger Bank sand. In *Frontiers in Offshore Geotechnics III, Proceedings of the 3rd International Symposium on Frontiers in Offshore Geotechnics, Oslo, Norway, 10–12 June 2015*; Meyer, V., Ed.; CRC Press: London, UK, 2015; pp. 1167–1172.
39. Blaker, Ø.; Andersen, K.H. Cyclic properties of dense to very dense silica sand. *Soils Found.* **2019**, *59*, 982–1000.
40. Cotterill, C.; Dove, D.; Long, D.; James, L.; Duffy, C.; Mulley, S.; Forsberg, C.F.; Tjelta, T.I. Dogger Bank-A Geo Challenge. In Proceedings of the Seventh International Conference, London, UK, 12–14 September 2012.
41. Forchheimer, P. Wasserbewegung durch boden. *Z. Ver. Deutsch, Ing.* **1901**, *45*, 1782–1788.
42. Chai, Z.; Shi, B.; Lu, J.; Guo, Z. Non-Darcy flow in disordered porous media: A lattice Boltzmann study. *Comput. Fluids* **2010**, *39*, 2069–2077. [CrossRef]
43. Sharqawy, M.H.; Lienhard, J.H.; Zubair, S.M. Thermophysical properties of seawater: A review of existing correlations and data. *Desal. Water Treat.* **2010**, *16*, 354–380.
44. Grossmann, S.; Lohse, D. Scaling in thermal convection: A unifying theory. *J. Fluid Mech.* **2000**, *407*, 27–56. [CrossRef]
45. Deng, W.; Carter, J.P. A theoretical study of the vertical uplift capacity of suction caissons. *Int. J. Offshore Polar Eng.* **2002**, *12*, 89–97.
46. Han, F.; Ganju, E.; Salgado, R.; Prezzi, M. Effects of Interface Roughness, Particle Geometry, and Gradation on the Sand–Steel Interface Friction Angle. *J. Geotech. Geoenvironmental Eng.* **2018**, *144*, 04018096.
47. Marek, K. Experimental investigation of skirted foundation in sand subjected to rapid uplift. *Arch. Hydro-Eng. Environ. Mech.* **2020**, *67*, 17–34.
48. Sgardeli, C.G. A Finite Element Analysis of the Pullout Capacity of Suction Caissons in Clay. Ph.D. Thesis, Massachusetts Institute of Technology, Cambridge, MA, USA, 2009.
49. Koh, K.X.; Wang, D.; Hossain, M.S. Numerical simulation of caisson installation and dissipation in kaolin clay and calcareous silt. *Bull. Eng. Geol. Environ.* **2017**, *77*, 953–962.
50. Sun, L.; Qi, Y.; Feng, X.; Liu, Z. Tensile capacity of offshore bucket foundations in clay. *Ocean Eng.* **2020**, *197*, 106893.
51. Idriss, I.M.; Boulanger, R.W. 2nd Ishihara lecture: SPT-and CPT-Based relationships for the residual shear strength of liquefied soils. *Soil Dyn. Earthq. Eng.* **2015**, *68*, 57–68.
52. Guo, Z.; Jeng, D.S.; Zhao, H.Y.; Guo, W.; Wang, L.Z. Effect of seepage flow on sediment incipient motion around a free spanning pipeline. *Coast. Eng.* **2019**, *143*, 50–62.

MDPI AG
Grosspeteranlage 5
4052 Basel
Switzerland
Tel.: +41 61 683 77 34

Sustainability Editorial Office
E-mail: sustainability@mdpi.com
www.mdpi.com/journal/sustainability

www.ingramcontent.com/pod-product-compliance
Lightning Source LLC
Chambersburg PA
CBHW061610210926
51850CB00009B/2277
9783725877713